T0282437

SpringerBriefs in Environmental Science

SpringerBriefs in Environmental Science present concise summaries of cutting-edge research and practical applications across a wide spectrum of environmental fields, with fast turnaround time to publication. Featuring compact volumes of 50 to 125 pages, the series covers a range of content from professional to academic. Monographs of new material are considered for the SpringerBriefs in Environmental Science series.

Typical topics might include: a timely report of state-of-the-art analytical techniques, a bridge between new research results, as published in journal articles and a contextual literature review, a snapshot of a hot or emerging topic, an in-depth case study or technical example, a presentation of core concepts that students must understand in order to make independent contributions, best practices or protocols to be followed, a series of short case studies/debates highlighting a specific angle.

SpringerBriefs in Environmental Science allow authors to present their ideas and readers to absorb them with minimal time investment. Both solicited and unsolicited manuscripts are considered for publication.

More information about this series at http://www.springer.com/series/8868

Theodoros Zachariadis

Climate Change in Cyprus

Review of the Impacts and Outline
of an Adaptation Strategy

 Springer

Theodoros Zachariadis
Cyprus University of Technology
Limassol
Cyprus

ISSN 2191-5547   ISSN 2191-5555   (electronic)
SpringerBriefs in Environmental Science
ISBN 978-3-319-29687-6   ISBN 978-3-319-29688-3   (eBook)
DOI 10.1007/978-3-319-29688-3

Library of Congress Control Number: 2016930280

Printed on acid-free paper

This Springer imprint is published by SpringerNature
The registered company is Springer International Publishing AG Switzerland

Contents

Chapter 1
Introduction

Abstract According to the current scientific consensus, warming of the global climate system seems to be unambiguous and is most likely due to anthropogenic emissions of greenhouse gases. Anthropogenic climate change has been characterised as 'the largest market failure' that mankind has ever been faced with. Mediterranean Europe is expected to experience the most adverse climate change effects compared to other European regions. Cyprus, an island state in the south-east Mediterranean, already has a semi-arid climate, with hot summers and mild winters, and the most severe water scarcity problem in Europe. The island is situated on a hot spot and is projected to face significant temperature increases and decline in precipitation levels. This book addresses in a complete yet concise manner the knowledge available by the end of 2015 about expected impacts from climate change in Cyprus and outlines the main ingredients of an adaptation strategy.

Keywords Climate adaptation · Climate impacts · Greenhouse gases · South-east Mediterranean

According to the current scientific consensus, warming of the global climate system seems to be unambiguous and is most likely due to anthropogenic emissions of greenhouse gases (IPCC 2014). Anthropogenic climate change has been characterised as 'the largest market failure' that mankind has ever been faced with (Stern 2007). Climate science has made considerable progress in recent years, and the science of natural and economic impacts from climate change has also experienced noteworthy advances. There are still areas with moderate to large uncertainty; however, because of the long-term nature of the problem, governments, enterprises and citizens have to be aware of the latest scientific insights in order to prepare for and adapt to potentially adverse impacts on their welfare.

© The Author(s) 2016
T. Zachariadis, *Climate Change in Cyprus*,
SpringerBriefs in Environmental Science, DOI 10.1007/978-3-319-29688-3_1

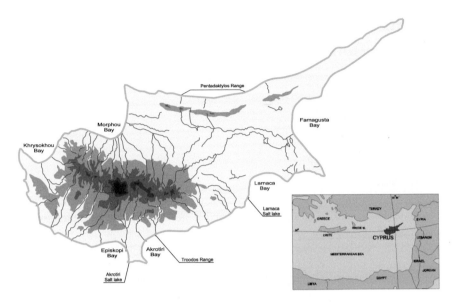

Fig. 1.1 Location and topography of Cyprus. *Source* Zoumides (2014)

Mediterranean Europe is expected to experience the most adverse climate change effects compared to other European regions. As shown in Fig. 1.1, Cyprus is an island state that is located in the south-east Mediterranean. It has a population of about 900,000 inhabitants and became a member of the European Union (EU) in 2004.[1] It already has a semi-arid climate, with hot summers and mild winters, and the most severe water scarcity problem in Europe (Eurostat 2015). With a wide-ranging variety of soil types (illustrated in Fig. 1.2) and very rich biodiversity, as will be explained in Chap. 3, the island is situated on a hot spot and is projected to face significant temperature increases and decline in precipitation levels. Irrespective of the greenhouse gas emission abatement efforts of the country, which are in line with the EU's decarbonisation strategy, serious negative effects of climate change should be expected in the coming decades in various sectors. This book attempts to address in a complete yet concise manner the knowledge available by the end of 2015 about expected impacts from climate change in Cyprus and outlines the main ingredients of an adaptation strategy.

Adverse effects, however, are not inevitable. Coping with changes caused by climate alterations is possible, provided that proactive actions are taken by both the public and the private sector. Therefore, this book provides a detailed list of adaptation measures by sector. Public authorities need to set clear priorities and implement well-designed policies in order to mitigate the main adverse impacts outlined in the report.

---

[1]The information provided here refers only to the area controlled by the government of the Republic of Cyprus.

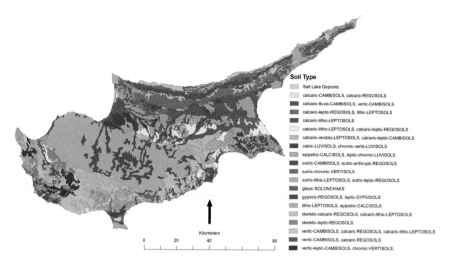

Fig. 1.2 Soil map of Cyprus. *Source* Hadjiparaskevas (2005)

The contents of this book are based on a review of the major recent studies on climate change, including studies of international, European and Mediterranean institutions and scientific articles, and—where available—specialised studies for the island of Cyprus. We first conducted such a review in years 2010–2011, in the frame of a research project funded by the Republic of Cyprus and the European Regional Development Fund through a grant from the Research Promotion Foundation of Cyprus (Shoukri and Zachariadis 2012). Since then, new knowledge and findings have appeared for the Mediterranean region and—to a more limited extent—specifically for Cyprus. Moreover, a National Adaptation Strategy for the Republic of Cyprus has been prepared by environmental authorities in the frame of another project that was cofunded by the European Union (CYPADAPT 2013). The review included in this book draws on those studies as well, citing clearly its data sources in each case.

Elpida Shoukri and Valia Constantinidou have assisted in collecting and presenting this material at various stages; I am grateful to both of them. The book has also benefited from information gathered through personal communication of the authors with public authorities, non-governmental organisations and researchers. I am thankful to a large number of people who have put time and effort to share their knowledge with us. Part of this work was funded by a Jean Monnet Module entitled '3EP—European Energy and Environmental Policy at a Crossroads', which was funded by the European Commission in the frame of its Lifelong Learning Programme. Some additional funding for preparing this book was provided in 2015 by own funds of the Cyprus University of Technology; both funding sources are gratefully acknowledged, as well as the editorial assistance of Isidoros Ziogou.

References

CYPADAPT. (2013). Report on the future climate change impact, vulnerability and adaptation assessment for the case of Cyprus. Deliverable 3.4, project CYPADAPT LIFE10 ENV/CY/000723. Available at http://uest.ntua.gr/cypadapt/wp-content/uploads/DELIVERABLE3.4.pdf

Eurostat. (2015). Online information on the Water Exploitation Index. Available at http://ec.europa.eu/eurostat/en/web/products-datasets/-/TSDNR310

Hadjiparaskevas, C. (2005). Soil survey and monitoring in Cyprus. *European Soil Bureau-Research Report, 9*, 97–101. Available at http://eusoils.jrc.ec.europa.eu/esdb_archive/eusoils_docs/esb_rr/n09_soilresources_of_europe/Cyprus.pdf

IPCC. (2014). Climate change 2014: Synthesis report. In Core Writing Team, R. K. Pachauri, & L. A. Meyer (Eds.), Contribution of Working Groups I, II and III to the Fifth Assessment Report of the Intergovernmental Panel on Climate Change (151 pp.). Geneva, Switzerland: IPCC. ISBN 978-92-9169-143-2

Shoukri, E., & Zachariadis, T. (2012). Climate change in Cyprus: Impacts and adaptation policies. Environmental Policy Research Group Report 01-12, Cyprus University of Technology, Limassol, Cyprus. Available at http://works.bepress.com/theodoros_zachariadis/24

Stern, N. (2007). *The economics of climate change: The Stern review*. Cambridge: Cambridge University Press.

Zoumides, C. (2014) Quantitative methods and tools towards sustainable agricultural water management in Cyprus. Ph.D. Dissertation, Cyprus University of Technology, Limassol, Cyprus.

Chapter 2
Facts and Projections on Climate Change

Abstract This chapter outlines the main facts about changes in the global climate in recent decades and established projections of climate scientists regarding the future evolution of worldwide greenhouse gas emissions, temperature, sea level and precipitation. Apart from the global level, the past and future evolution of key climate variables in Europe, the Mediterranean region and Cyprus is described. Latest available information from international and regional studies is used for this purpose.

Keywords Climate model · Concentration pathways · Precipitation · Radiative forcing · Sea level · Temperature

2.1 Definition of Climate Change, Main Facts and Major Projections

2.1.1 Definitions of Climate Change

According to the official definition given by the United Nations Intergovernmental Panel on Climate Change (IPCC), *climate change* refers to a statistically significant variation in either the mean state of the climate or its variability, persisting for an extended period (typically decades or longer). Climate change may be due to natural internal processes or external causes, or to persistent anthropogenic changes in the composition of the atmosphere or in land use (IPCC 2007a). The United Nations Framework Convention on Climate Change (UNFCCC) focuses specifically on anthropogenic climate change, i.e. '*the change of climate that is attributed directly or indirectly to human activity, that alters the composition of the global atmosphere and that is in addition to natural climate variability observed over comparable time periods*' (UN 1992).

© The Author(s) 2016
T. Zachariadis, *Climate Change in Cyprus*,
SpringerBriefs in Environmental Science, DOI 10.1007/978-3-319-29688-3_2

Some additional definitions associated with climate change that have been addressed by the UNFCCC (1992) are presented below and aim at providing a better understanding of the overall situation:

Climate system means the totality of the atmosphere, hydrosphere, biosphere and geosphere and their interactions.

Adverse effects of climate change are changes in the physical environment or biota resulting from climate change which have significant deleterious effects on the composition, resilience or productivity of natural and managed ecosystems or on the operation of socio-economic systems or on human health and welfare.

Greenhouse gases (GHGs) are those gaseous constituents of the atmosphere, both natural and anthropogenic, that absorb and re-emit infrared radiation. Their total amount is usually expressed in tonnes of equivalent carbon dioxide (CO_{2eq}), a metric that takes into account the global warming potential of each gas. The IPCC considers the following GHGs: carbon dioxide (CO_2), methane (CH_4), nitrous oxide (N_2O), hydrofluorocarbons (HFCs), perfluorocarbons (PFCs) and sulphur hexafluoride (SF_6).

Emissions denote the release of greenhouse gases and/or their precursors into the atmosphere over a specified area and period of time.

It is also important to note some additional definitions that are useful to understand fully the aspects of climate change:

Macroclimate: the overall climate of a region (usually a large geographical area).

Microclimate: the essentially uniform local climate of a usually small site or habitat.[1]

2.1.2 Main Facts

The vast majority of studies released in the past decade mention clearly that anthropogenic GHG emissions have been increasing substantially in the last decades, thus contributing to large-scale climate change. Human influence on the climate system is clear, and recent anthropogenic emissions of greenhouse gases are the highest in history. About half of the cumulative anthropogenic CO_2 emissions between 1750 and 2010 have occurred in the last 40 years.

As shown in Fig. 2.1, greenhouse gas emissions have been increasing from 1970 to 2010, mainly because of man-caused activities. It is widely accepted that total anthropogenic GHG emissions were the highest in human history from 2000 to 2010, reaching 49 ($\pm$4.5) billion tonnes (Gt) CO_{2eq} per year in 2010. Of those, CO_2 accounted for 76; 16 % came from CH_4, 6 % from N_2O and 2 % from fluorinated gases.

---

[1]Definitions of macro- and microclimate according to Merriam–Webster's dictionary, available online at http://www.merriam-webster.com/dictionary/, last accessed in December 2015.

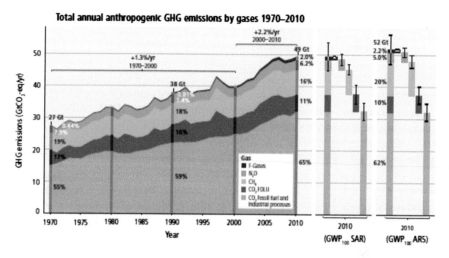

Fig. 2.1 Evolution of anthropogenic GHG emissions since 1970. *Source* IPCC (2014)

On a global scale, economic growth and population growth are the most important reasons for increasing fossil fuel combustion-related CO_2 emissions. Whilst the contribution of population growth has remained roughly the same in the last four decades, the contribution of economic growth to emission growth has risen sharply during the last years. Between 2000 and 2010, both drivers outpaced emission reductions caused by improvements in energy intensity. Moreover, increased use of coal relative to other energy sources has reversed the long-standing trend of gradual decarbonisation of the world's energy supply.

According to the IPCC 5th Assessment Report (or AR5), released in 2014, warming of the climate system is unequivocal, and since the 1950s, many of the observed changes are unprecedented over decades to millennia. The atmosphere and ocean have warmed, the amounts of snow and ice have receded, sea level has risen, and the concentrations of greenhouse gases have increased. New atmospheric temperature measurements show an estimated warming of 0.85 °C since 1880, with the fastest rate of warming in the Arctic (IPCC 2014).

In addition, it is 'virtually certain' that the upper 700 m of the Earth's oceans has warmed during the period from 1971 and 2010. The deep ocean, below 3000 m in depth, 'likely' warmed between 1992 and 2005. Moreover, it can be stated with 'high confidence' that glaciers have shrunk worldwide and that the Greenland and Antarctic ice sheets have lost mass over the past two decades. The report notes with 'very high confidence' that ice loss from Greenland has accelerated during the past two decades. Greenland is now losing about 215 Gt per year of ice, whilst the rest of the world's glaciers lose about 226 Gt per year.

In summary, the conclusion that much of the global warming over the past fifty years is due to human activities is considered *extremely likely* in the IPCC AR5 report, upgraded from *very likely* in the previous IPCC report (4th Assessment Report or AR4; IPCC 2007a).

2.1.3 Projections of GHG Concentrations and Radiative Forcing

IPCC AR5 reports that without additional efforts to reduce GHG emissions, emission growth is expected to persist throughout the twenty-first century. Baseline scenarios, i.e. those without additional emission abatement efforts, forecast GHG concentrations in the atmosphere to exceed 450 parts per million (ppm) CO_{2eq} by 2030 and reach concentration levels between 750 and more than 1300 ppm CO_{2eq} by the year 2100. For comparison, the CO_{2eq} concentration in 2011 is estimated to be 430 ppm (uncertainty range 340–520 ppm). As a result, global mean surface temperature in the 2100 is projected to increase by 3.7–4.8 °C compared to preindustrial levels (IPCC 2014). A set of four scenarios of 'representative concentration pathways' (RCPs) has been the basis of these projections and for the assessment of their global impacts. Derived from a very large number of GHG emission projections collected from the international literature, these RCPs assume radiative forcing levels by the year 2100 of 2.6, 4.5, 6.0 and 8.5 watts per square metre (W/m^2), corresponding to GHG concentrations of 450, 650, 850 and 1370 ppm CO_{2eq}, respectively, covering the range of potential anthropogenic climate forcing in the twenty-first century as reported in the literature. Figure 2.2 illustrates the different scenarios considered in that study.

This kind of scenarios differs somewhat from the approach followed by IPCC in the past. Figure 2.3 compares the four RCPs mentioned above with the modest emission scenario A1B that was used by (IPCC 2007a), which is the basis for many of the climate change impact studies that will be cited in this book. Evidently, the choice of the scenario is not very important until the mid-twenty-first century; for later years, A1B seems to lie between the two most 'pessimistic' (i.e. highest) RCPs of IPCC AR5.

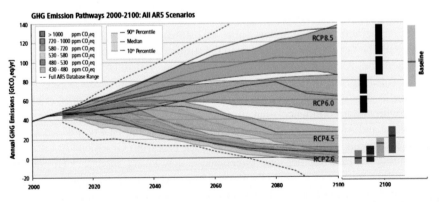

Fig. 2.2 The range of scenarios that were considered in the IPCC 5th Assessment Report. *Source* IPCC (2014)

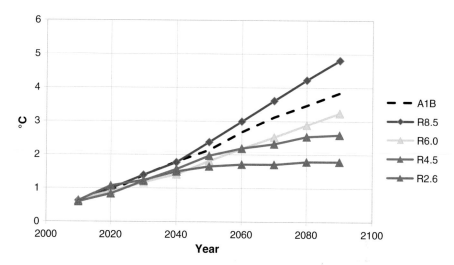

Fig. 2.3 Comparison of predicted changes in global mean surface temperature (compared to the 1986–2005 period) according to the modest A1B emission scenario of IPCC (2007a) and the four different RCPs developed by IPCC (2014). *Source* Panos Hadjinicolaou, The Cyprus Institute, Nicosia, Cyprus (personal communication, 2015)

GHG mitigation scenarios in which it is likely that the temperature increase caused by anthropogenic GHG emissions can be kept to less than 2 °C relative to preindustrial levels (which is a widely stated policy objective) are characterised by atmospheric concentrations in 2100 of about 450 ppm CO_{2eq}. This can only be achieved through substantial cuts in anthropogenic GHG emissions by mid-century, thanks to large-scale changes in energy systems including serious energy efficiency improvements and a substantial shift to low- or zero-carbon energy systems. Scenarios reaching these concentrations by 2100 are characterised by lower global GHG emissions in 2050 than in 2010 and near-zero emission levels by the end of the century.

At the other end of the projections, 'business as usual' emission scenarios lead to concentrations that may exceed 650 ppm CO_{2eq} by 2100, and these are very unlikely to limit temperature change to below 2 °C relative to preindustrial levels. This is confirmed by many global emission forecasts. For example, the Organisation for Economic Co-operation and Development (OECD) *Environmental Outlook* baseline scenario envisaged that without more ambitious policies than those in force today, GHG emissions would increase by another 50 % by 2050, primarily driven by a projected 70 % growth in CO_2 emissions from energy use due to an 80 % increase in global energy demand (OECD 2015). More recently, both the International Energy Agency and the OECD, shortly before the start of the international climate summit in Paris in November 2015, stated that national carbon emission mitigation targets and policies up to that date were clearly insufficient to reach the 2 °C target and that stronger actions were necessary in order to remain on track for a '2 °C world' (IEA 2015; OECD 2015).

2.2 Global Projections

2.2.1 Temperature

Future climate will depend both on warming caused by already existing green-house gases in the atmosphere due to past anthropogenic emissions and on future anthropogenic emissions and natural climate variability. The upper part of Fig. 2.4 illustrates the main temperature projections of the IPCC AR5. Assuming that there will be no major event such as volcanic eruptions, changes in solar irradiance or changes in natural sources of GHG emissions, global mean surface temperature change for the period 2016–2035 relative to 1986–2005 is similar for the four RCPs and will likely be in the range of 0.3–0.7 °C. It is virtually certain that there will be more frequent hot and fewer cold temperature extremes over most land areas on daily and seasonal timescales, and it is very likely that heatwaves will occur with a higher frequency and longer duration.

Relative to 1850–1900, global surface temperature change for the end of the twenty-first century (2081–2100) is projected to likely exceed 1.5 °C for RCP4.5, RCP6.0 and RCP8.5 scenarios. Warming is unlikely to exceed 2 °C only for RCP2.6. Moreover, the Arctic region is expected to continue to warm more rapidly than the global mean.

2.2.2 Sea-Level Rise

Recent IPCC AR5 projections of sea-level rise are displayed in the lower part of Fig. 2.4. Global sea level is rising mainly because of:

- thermal expansion of warming ocean water;
- addition of new water from ice sheets of Greenland and Antarctica, from gla-ciers and ice caps;
- addition of water from land surface run-off (IPCC 2007b).

The new projections show an increase of 0.26–0.55 m by 2100 under a low-emis-sion scenario and 0.52–0.98 m under the high-emission scenario. By contrast, the previous IPCC report (AR4) did not include some of the effects of ice sheet move-ment due to warming and therefore published much lower estimates in the range of 0.18–0.38 and 0.26–0.59 m under a low- and high-emission scenario, respec-tively, by 2100.

It is also likely that the Arctic Ocean will be ice-free during part of the summer before 2050 under a high-emission scenario. This represents a large change from

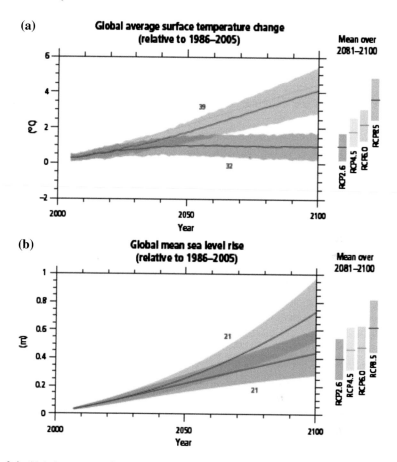

Fig. 2.4 Global average surface temperature change and global mean sea-level rise according to the four RCPs of IPCC (2014)

the AR4, which estimated that the Arctic Ocean would not be ice-free during the summer until late in the twenty-first century. The AR5 finds that Arctic sea ice surface extent has decreased by 3.5–4.1 % per decade (9.4–13.6 % during summer), which is higher than the corresponding AR4 estimate.

2.2.3 Precipitation

Changes in precipitation will not be uniform. The high latitudes and the equatorial Pacific are likely to experience an increase in annual mean precipitation under the RCP8.5 scenario. In many mid-latitude and subtropical dry regions, mean precipitation is expected to decrease, whilst in many mid-latitude wet regions, mean precipitation will likely increase (RCP8.5 scenario). Extreme precipitation events over most of the mid-latitude land masses and over wet tropical regions will very likely become more intense and more frequent (IPCC 2014).

2.3 Projections for Europe and the Mediterranean

2.3.1 Past Trends and Projections for Europe

There is a significant agreement amongst recent studies that all emission scenarios will lead to higher temperatures all over Europe, with strongest warming projected in Southern Europe in summer and in Northern Europe in winter (JRC 2014; Kjellström et al. 2011). Since the 1980s, warming has been strongest over Scandinavia, especially in winter, whereas the Iberian Peninsula warmed mostly in summer (EEA 2012). The decadal average temperature over land area for 2002–2011 is 1.3 ± 0.11 °C above the 1850–1899 average.

Although many features of the simulated climate change in Europe and the Mediterranean are qualitatively consistent amongst models, there are uncertainties that reflect the sensitivity of the European climate to the magnitude of global warming and the changes in atmospheric circulation. The main uncertainty is the substantial natural variability of European climate, particularly for short-term climate projections (IPCC 2007a). The most important projections for Europe are summarised below (EEA 2012; IPCC 2014):

- Temperature increases from 1.0 to 5.5 °C are expected by 2100, which is higher than the projected global warming (1.8–4.0 °C), with the largest warming over Eastern and Northern Europe in winter and over Southern Europe and Mediterranean in summer.
- Changes in precipitation show more spatially variable trends across Europe. Annual precipitation has increased in Northern Europe (up to +70 mm per decade) and decreased in parts of Southern Europe.
- The intensity of precipitation extremes has increased in the past 50 years, and these events are projected to become more frequent.
- High-temperature extremes (hot days, tropical nights and heatwaves) have become more frequent, whilst low-temperature extremes (cold spells, frost days) have become less frequent.
- Winter snow cover has a high interannual variability and an insignificant negative trend over the period 1967–2007 (Henderson and Leathers 2010).
- Europe is marked by increasing mean sea level with regional variations, except in the northern Baltic Sea, where the relative sea level decreased due to vertical crustal motion (Menéndez and Woodworth 2010; Albrecht et al. 2011; EEA 2012).
- Mean wind speeds have declined over Europe over recent decades (Vautard et al. 2010)—although this finding should be treated with caution because of problematic anemometer data and climate variability.

2.3.2 Projections for the Mediterranean

The Mediterranean Sea region, especially the southern and eastern rim, has been identified as one of the main climate change hot spots (i.e. one of the areas most sensitive to climate change) in the world due to water scarcity, concentration of economic activities in coastal areas and reliance on climate-sensitive agriculture. However, the region itself emits low levels of greenhouse gases compared to other areas in the world. CO_2 emission data show that in 2009, the Mediterranean countries together emitted 6.7 % of the world's emissions, equivalent to more than 2 billion tonnes of CO_2. This amount has increased by a factor of 4 in the last 50 years, with an increase in the contribution from countries from the southern region of the Mediterranean from 9 to 30 %. Meanwhile, the contribution of all EU Mediterranean countries has decreased over the same period from 88 to 54 %.

Satellite altimetry data show that the mean sea level of the Mediterranean increased by 2.6 cm in the period 1992–2008. This change seems to be lower than global sea-level rise, which has reached a rate of 3.0–3.5 mm/year since 1992. This different behaviour indicates that being a semi-enclosed basin, the Mediterranean does not respond linearly to the influences of the open ocean on the timescale in question (Vigoa et al. 2011).

The Mediterranean region lies in a transition zone between the arid climate of North Africa and the temperate and rainy climate of Central Europe and is affected by interactions between mid-latitude and tropical processes. Because of these features, even relatively minor modifications of the general circulation in the atmosphere can lead to substantial changes in the Mediterranean climate. This makes the Mediterranean a potentially vulnerable region to climatic changes. In the last decades in the Mediterranean region, temperatures have risen faster than the global average and model projections agree on its future warming and drying, with a likely increase of heatwaves and dry spells, especially since it is expected to experience high temperature increases, reduced precipitation and more frequent droughts—which adds to the already existing water scarcity (Lionello et al. 2014).

Based on the latest comprehensive climate change vulnerability assessment carried out by the European Environment Agency (EEA 2012), the Mediterranean region is projected to experience:

- Temperature increases that will be larger than the European average;
- Lower annual precipitation levels and decreased annual river flows;
- Increased risks of biodiversity loss and desertification;
- Adverse effects on forest fires, summer tourism, agricultural production and public health.

Collaborative research projects have attempted to describe in detail how the climate is expected to change in Europe under various emission scenarios. For example, by using regional climate model simulations, the ENSEMBLES project evaluated many indicators. Projections of the ENSEMBLES project on changes in

Fig. 2.5 Projected changes in annual mean surface air temperature under the IPCC AR4 A1B scenario, 2021–2050 (*left*) and 2071–2100 (*right*). *Source* van der Linden and Mitchell (2009)

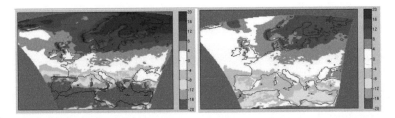

Fig. 2.6 Projected changes in annual precipitation under the IPCC AR4 A1B scenario, 2021–2050 (*left*) and 2071–2100 (*right*). *Source* van der Linden and Mitchell (2009)

annual mean surface air temperature and annual precipitation over Europe are presented in Figs. 2.5 and 2.6, respectively.[2]

The PESETA project (Projection of Economic Impacts of Climate Change in Sectors of the European Union based on Bottom-Up Analysis) was a major project carried out mainly by the Joint Research Centre of the European Commission. In its first phase (JRC 2009), it examined three important indicators—temperature, precipitation and sea-level rise—and their projected evolution up to the 2071–2100 period based on the IPCC AR4 scenarios A2 and B2, by using data from several standardised high-resolution climate projections. It was clear from the project results that the most adverse effects around Europe (in terms of temperature increase and fall in precipitation) are expected to occur in South Europe.

The PESETA II project (JRC 2014) updated and expanded the previous analysis. Distinguishing between five regions (Northern Europe, UK and Ireland, Central Europe, Central Europe South and Southern Europe), it employed a bottom-up approach to analyse ten biophysical impact categories (agriculture, energy, river floods, droughts, forest fires, transport infrastructure, coasts, tourism, habitat suitability of forest tree species and human health) considering a broad range of climate model simulations. Results for most of these impact categories were

---

[2]Graphs and results reproduced here are from the EU-funded project ENSEMBLES (Contract number 2003-505539) (van der Linden and Mitchell 2009)—more details can be found on the website www.ensembles-eu.org.

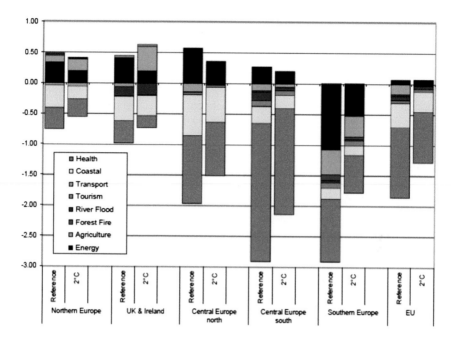

Fig. 2.7 Projected impact on economic welfare (as a percentage of GDP) for two climate change scenarios according to results of the PESETA II project (JRC 2014)

then entered in an economic model to assess economic and welfare effects of the climate scenarios. Impacts were evaluated for three emission scenarios covering a range of low-emission to high-emission pathways.

Results indicate that, for a 'reference' scenario without serious mitigation of greenhouse gas emissions, annual climate-related damage around Europe may amount to 190 billion euros or 2 % of EU's gross domestic product (GDP). The most interesting finding is that the geographical distribution of these climate damages is asymmetric, with a clear bias towards Southern European regions. As shown in Fig. 2.7, welfare losses range from 0.2 % of GDP in Northern Europe to 3 % of GDP in the Central Europe South and Southern European regions, i.e. fifteen times higher than the damage in Northern Europe. The highest welfare losses will occur in Southern Europe (€74 billion) and Central Europe South (€58 billion)—in fact, the damage in the two southern regions will account for more than two-thirds of total EU-wide costs. Most of the economic losses are expected to come from increased mortality, coastal damages and costs of reduced agricultural production.

The MedCLIVAR project (Lionello 2012) has also focused on the Mediterranean's climate conditions and future changes. In the frame of that project, Dubrovský et al. (2014) have used results from sixteen global circulation models and found that most models agree that there will be an increase in temperature in all seasons and for all parts of the Mediterranean by the end of the twenty-first century. Precipitation is projected to decrease in almost all parts of the region and in all seasons, but especially in summer. A significant decline of soil moisture is also forecast, with the most significant decrease occurring in summer. Temperature maxima will increase not only because of an overall rise in mean temperature, but to some extent also because of increases in temperature variability and daily temperature range. Higher daily precipitation extremes and longer drought spells are expected all across the region. Regional climate models for the Mediterranean, obtained through 'downscaling' of global circulation models, also agree in terms of temperature increases—although there are more discrepancies amongst models as regards projections of precipitation and weather extremes (Jacobeit et al. 2014).

As regards especially the Eastern Mediterranean, two recent studies by Greek and Israeli authorities, based on the IPCC AR4 scenarios, provide valuable information that is of interest to Cyprus.

The report 'Environmental, economic and social impacts of climate change in Greece', prepared for the Bank of Greece (2011), assesses the main climatic indicators and presents projections on many important parameters for the Greek territory. According to its findings:

- The mean annual air temperature for the periods of 2021–2050 and 2071–2100 in relation to the reference period 1961–1990 will increase in the entire territory of Greece.
- Precipitation will decrease, with a more intense reduction for the A2 and A1B emission scenarios.
- Mean sea-level rise for Greece has been based on IPCC projections, but the study adds the importance of other factors for the estimation of the vulnerability of a coastal area to sea-level rise, such as tectonics, sediments and coastal geomorphology.

Furthermore, Israel's Second National Communication on Climate Change (Israel's Ministry of Environmental Protection 2010) that was prepared for the UNFCCC forecasts:

- A seasonal rise in air temperature of 1.5–2.0 °C in the south-eastern Mediterranean over a thirty-year period.
- An average temperature increase of 5 and 3.5 °C according to projections based on emission scenarios A2 and B2, respectively, in the years 2071–2100, compared to 1961–1990.
- A 10 % decrease in precipitation in Israel by 2020, reaching a 20 % decrease by 2050.
- An increase in the number of days with extreme temperatures and in the number of extreme rainfall events, along with a decrease in the amount of seasonal rain.
- Sea-level rise in the Israeli waters of about 0.5 m by 2050 and approximately 1 m by 2100.

Other climate modelling studies that focused on the Eastern Mediterranean and are of direct relevance for Cyprus are discussed in Sect. 2.4.2.

2.4 Past Trends and Projections for Cyprus

2.4.1 Trends in Major Climate Variables

At a latitude 35° north and a longitude 33° east, Cyprus has an intense Mediterranean climate, with hot dry summers from mid-May to mid-September and rainy, rather changeable, winters from November to mid-March. The meteorology of the island is significantly affected by the two mountainous areas shown in Fig. 1.1: The central Troodos massif, rising to 1951 m, and the long narrow Kyrenia mountain range, with peaks of about 1000 m. The predominantly clear skies and high amounts of sunshine (a typical year includes more than 300 days of sunshine) give rise to large seasonal and daily differences between temperatures of the sea and the interior of the island, which also cause considerable local weather effects. In summer months of July and August, the mean daily temperature ranges between 29 °C on the central plain and 22 °C on the Troodos mountain, whilst the average maximum temperature for these months ranges between 36 and 27 °C, respectively. In January, the mean daily temperature is 10 °C on the central plain and 3 °C on the highest parts of the Troodos mountain, with an average minimum temperature of 5 and 0 °C, respectively (CMS 2015).

Observations from the beginning of the twentieth century show an increasing trend in the annual mean temperature in Cyprus, with a rate of increase of 0.01 °C per year. Overall, a warming of approximately 1–1.58 °C has been observed over the twentieth century. This increase exceeds the mean global temperature rise observed for the same period. The rates of change of precipitation and temperature have been greater during the second half of the twentieth century. According to the Cyprus Department of Meteorology,[3] most of the warm years in the century have been recorded after 1990.

As illustrated in Fig. 2.8, a temperature increase has been recorded both in towns and in rural areas of the country. The stronger urban heat island effect played an important role in the temperature increase in towns; however, the increase in temperature in rural areas is indicative of the regional or global climate changes during the last decades. This finding is reinforced by satellite remote sensing data and in situ measurements of sea surface temperatures, which indicate that a general warming has occurred in the Levantine basin (the easternmost part of the Mediterranean Sea) over the period 1996–2006, both at interannual and at seasonal timescales (Samuel-Rhoads et al. 2009).

Moreover, changes in the diurnal temperature range have been recorded during the twentieth century; minimum daily temperatures have generally increased more than the maximum daily temperatures, resulting in a decrease in the long-term diurnal temperature range. This decrease is very similar to some of the coastal

---

[3]See the Department's web page (http://www.moa.gov.cy/ms) for graphs and brief reports about the climate of Cyprus.

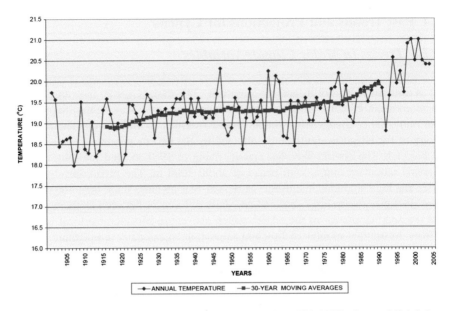

Fig. 2.8 Evolution of annual mean temperature in Nicosia (1901–2005). *Source* Official data cited by Shoukri and Zachariadis (2012)

stations from southern Israel, which show higher minimum temperatures but lower maximum temperatures (Price et al. 1999).

Figure 2.9 displays the evolution of mean annual precipitation in the country, which demonstrated a decreasing trend over the twentieth century. The rate of decrease was 1 mm/year on average. The decrease in mean precipitation was larger during the second half of the twentieth century because years with low precipitation or drought occurred more frequently in comparison with those in the first half of the century: average annual precipitation in the first 30-year period of the twentieth century was 559 mm, whilst average precipitation in the last 30-year period was 462 mm, which corresponds to a decrease of 17 %.

No consistent data series are available for sea-level rise at the coasts of Cyprus. Vertical land movement/tectonics must be taken into consideration, amongst other parameters, for the estimation of sea-level rise for the Cyprus seas, since there is evidence that these parameters are counterbalancing any climate change-induced sea-level rise (Bank of Greece 2011).

2.4.2 Projections

Results of climate models were presented in Sect. 2.3 for Europe or the Mediterranean as a whole need to be refined in order to provide meaningful projections for the island of Cyprus. For several reasons, detailed regional climate

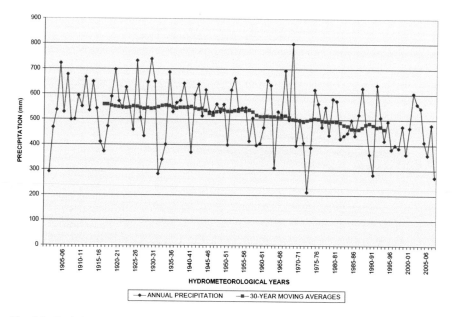

Fig. 2.9 Evolution of annual precipitation in Cyprus (1901–2006). *Source* Official data cited by Shoukri and Zachariadis (2012)

model (RCM) simulations are necessary instead of the more aggregate global circulation models (GCMs). The current horizontal resolution of GCMs used in century-long climate simulations is around 150 km, which does not resolve regional climate forcings associated with orography, coastlines and land surface properties. In order to adequately quantify potential impacts of climate change, climate projections are needed at a much finer resolution than that of GCMs—of a few tens of kilometres or less; this can be achieved through 'downscaling' methodologies. These can be classified as 'statistical downscaling' approaches, such as the ones reviewed by Jacobeit et al. (2014), or 'dynamic downscaling' methods like those mentioned below. Moreover, as Aufhammer et al. (2013) explain in detail, the use of GCMs for examining economic effects of climate change may lead to substantial errors due to aggregation bias; properly developed RCMs greatly improve the reliability of these impact assessments.

There have been some efforts applying different mesoscale models or RCMs to project the effects of anthropogenic global warming in the Eastern Mediterranean (EM) region by the end of this century. Önol and Semazzi (2010) concluded that temperature would increase by 4 °C and precipitation decrease by 20–30 %. Evans (2009) found a temperature increase of 2–4 °C in winter and 2–6 °C in summer. Precipitation decreased strongly (by 30–50 %) in winter along the EM coast. Lelieveld et al. (2012) presented a comprehensive regional climate assessment and discussed potential impacts of future climate change in the EM region based on projections from the PRECIS RCM. With emissions following the IPCC AR4 A1B

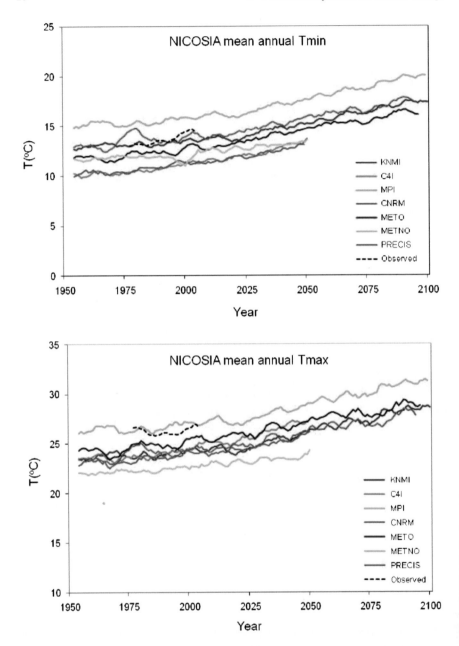

Fig. 2.10 Time series of mean annual minimum (Tmin) and maximum (Tmax) temperatures in Nicosia, according to observations and regional climate model projections. *Source* Giannakopoulos et al. (2012)

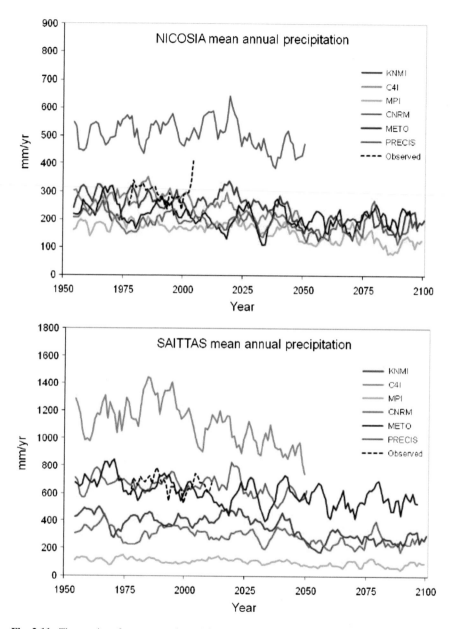

Fig. 2.11 Time series of mean annual precipitation in the city of Nicosia and a semi-mountainous area of Cyprus, according to observations and regional climate model projections. *Source* Giannakopoulos et al. (2012)

scenario, they found a gradual warming of about 3.5–7 °C throughout the twenty-first century compared to 1961–1990. They also demonstrated that this simulation was able to capture the mean climatic conditions as well as the increasing temperature tendency of the twentieth century in the region. For Cyprus, PRECIS projects:

- An almost linear warming throughout the twenty-first century up to +4 °C by the end of the century, with large interannual variability. Summer warming is projected to be approximately 1 °C larger than winter warming (Hadjinicolaou et al. 2011).
- Intense heat extremes. Lelieveld et al. (2014) analysed the PRECIS output for Nicosia (amongst other capitals in the EM region) and found that heatwave duration could increase by 7–10 times by 2099, exceeding 4 months per year, and that the coolest summers at the end of the century will be warmer than the hottest ones in the recent past.
- Rainfall decrease by 2–8 % (although this was not statistically significant).
- Lower precipitation frequency in the areas of Nicosia (inland) and Limassol (coast), whilst mountainous areas could experience a somewhat more frequent rainfall.
- An increase in the number of summer days by more than two weeks per year.
- A significant rise in the number of tropical nights—they are projected to increase by around one month per year.

Further simulations with PRECIS, followed by comparisons with six additional regional climate models used in the ENSEMBLES project mentioned in Sect. 2.3.2, reinforce these findings (Giannakopoulos et al. 2012). Figures 2.10 and 2.11 display the projected evolution of minimum and maximum temperatures and annual precipitation levels in the capital city of Nicosia, respectively, according to the different models employed by the authors. As has already been mentioned in Sect. 2.3, agreement between models is stronger in forecasts of temperature changes rather than in precipitation projections.

References

Albrecht, F., Wahl, T., Jensen, J., & Weisse, R. (2011). Detemining sea level change in the German Bight. *Ocean Dynamics, 61*, 2037–2050.

Aufhammer, M., Hsiang, S. M., Schlenker, W., & Sobel, A. (2013). Using weather data and climate model output in economic analyses of climate change. *Review of Environmental Economics and Policy, 7*, 181–198.

Bank of Greece. (2011). *Environmental, economic and social climate change impacts in Greece.* Athens, Greece, ISBN 978-960-7032-49-2

CMS (Cyprus Department of Meteorology). (2015). Graphs and statistics available at the Department's website http://www.moa.gov.cy/ms.

Dubrovský, M., Hayes, M., Duce, P., Trnka, M., Svoboda, M., & Zara, P. (2014). Multi-GCM projections of future drought and climate variability indicators for the Mediterranean region. *Regional Environmental Change 14*, 1907–1919.

EEA (European Environment Agency). (2012). Climate change, impacts and vulnerability in Europe 2012. EEA Report No. 12/2012, Copenhagen, Denmark. ISBN:978-92-9213-346-7, doi:10.2800/66071

Evans, J. P. (2009). 21st century climate change in the Middle East. *Climatic Change, 92*(3-4), 417–432.

Giannakopoulos, C., Petrakis, M., Kopania, T., Lemesios, G., & Roukounakis, N. (2012). Projection of climate change in Cyprus with the use of selected regional climate models. Deliverable 3.2 of CYPADAPT project funded by the EU Life Programme (project no. LIFE10 ENV/CY/000723). Available at http://cypadapt.uest.gr/?page_id=105

Hadjinicolaou, P., Giannakopoulos, C., Zerefos, C., Lange, M. A., Pashiardis, S., & Lelieveld, J. (2011). Mid-21st century climate and weather extremes in Cyprus as projected by six regional climate models. *Regional Environmental Change, 11*, 441–457.

Henderson, G. R., & Leathers, D. J. (2010). European snow cover extent variability and associations with atmospheric forcings. *International Journal of Climatology, 30*, 1440–1451.

IEA (International Energy Agency). (2015). *World energy outlook 2015.* Paris, France, ISBN:978-92-64-24366-8

IPCC. (2007a). *Synthesis report.* In R. K. Pachauri & A. Reisinger A. (Eds.), Contribution of Working Groups I, II and III to the Fourth Assessment Report of the Intergovernmental Panel on Climate Change. Geneva, Switzerland: IPCC, 104 pp.

IPCC. (2007b). *Climate change 2007: Impacts, adaptation and vulnerability.* In M. L. Parry, O. F. Canziani, J. P. Palutikof, P. J. Van der Linden & C. E. Hanson (Eds.), Contribution of Working Group II to the Fourth Assessment Report of the Intergovernmental Panel on Climate Change. Cambridge, UK: Cambridge University Press, pp. 315–356.

IPCC. (2014). *Climate change 2014: Synthesis report.* In R. K. Pachauri & L. A. Meyer (Eds.), Contribution of Working Groups I, II and III to the Fifth Assessment Report of the Intergovernmental Panel on Climate Change. Geneva, Switzerland: IPCC, 151 pp., ISBN 978-92-9169-143-2

Israel's Ministry of Environmental Protection. (2010). Israel's Second National Communication on Climate Change. Submitted under the United Nations Framework Convention on Climate Change.

Jacobeit, J., Hertig, E., Seubert, S., Lutz, K. (2014). Statistical downscaling for climate change projections in the Mediterranean region. *Regional Environmental Change 14*:1891–1906.

JRC (European Commission's Joint Research Centre). (2009). Climate change impacts in Europe—Final report of the PESETA research project. Projection of Economic Impacts of Climate Change in Sectors of Europe based on Bottom-up Analyses. Available at http://ipts.jrc.ec.europa.eu/publications/pub.cfm?id=2879

JRC (European Commission's Joint Research Centre). (2014). Climate impacts in Europe. The JRC PESETA II Project. JRC Scientific and Policy Reports, EUR 26586EN. Available at http://ipts.jrc.ec.europa.eu/publications/pub.cfm?id=7181

Kjellström, E., Nikulin, G., Hansson, U., Strandberg, G., & Ullerstig, A. (2011). 21st century changes in the European climate: Uncertainties derived from an ensemble of regional climate model simulations. *Tellus, 63A*, 24–40.

Lelieveld, J., Hadjinicolaou, P., Kostopoulou, E., Chenoweth, J., El Maayar, M., Giannakopoulos, C., et al. (2012). Climate change and impacts in the eastern Mediterranean and the Middle East. *Climatic Change, 114*, 667–687.

Lelieveld, J., Hadjinicolaou, P., Kostopoulou, E., Giannakopoulos, C., Pozzer, A., Tanarhte, M., & Tyrlis, E. (2014). Model projected heat extremes and air pollution in the Eastern Mediterranean and the Middle East in the 21st century. *Regional Environmental Change, 14*, 1937–1949.

Lionello, P. (Ed.). (2012). *The climate of the Mediterranean region—From the past to the future.* Elsevier. ISBN:9780124160422

Lionello, P., Abrantes, F., Gacic, M., Planton, S., Trigo, R., Ulbrich, U. (2014). The climate of the Mediterranean region: Research progress and climate change impacts. *Regional Environmental Change 14*, 1679–1684.

Menéndez, M., & Woodworth, P. L. (2010). Changes in extreme high water levels based on a global tide-gauge data set. *Journal of Geophysical Research: Oceans 115*(C10).

Önol, B., & Semazzi, D. (2010). Regionalization of climate change simulations over the eastern Mediterranean. *Journal of Climate, 22*, 1944–1961.

OECD (Organisation for Economic Cooperation and Development). (2015). *Climate change mitigation: Policies and progress.* France, Paris ISBN:978-92-64-23878-7

Price, C., Michaelides, S., Pashiardis, S., & Alpert, P. (1999). Long term changes in diurnal temperature range in Cyprus. *Atmospheric Research, 51*, 85–98.

Samuel-Rhoads, Y., Iona, S., Zodiatis, G., Hayes, D., Gertman, I., & Georgiou, G. (2009). Sea surface temperature and salinity variability in the Levantine Basin during the last decade, 1996 to 2006. Geophysical Research Abstracts, Vol. 11. General Assembly 2009.

Shoukri, E., & Zachariadis, T. (2012). Climate change in Cyprus: Impacts and adaptation policies. Environmental Policy Research Group Report 01-12, Cyprus University of Technology, Limassol, Cyprus. Available at http://works.bepress.com/theodoros_zachariadis/24

UN (United Nations). (1992). United Nations framework convention on climate change. Available at http://unfccc.int

Van der Linden, P., & Mitchell, J. F. B. (Eds.). (2009). ENSEMBLES: Climate change and its impacts: Summary of research and results from the ENSEMBLES project. Met Office Hadley Centre, FitzRoy Road, Exeter EX1 3 PB, UK, 160 pp.

Vautard, R., Cattiaux, J., Yiou, P., Thepaut, J. N., & Ciais, P. (2010). Northern Hemisphere atmospheric stilling partly attributed to an increase in surface roughness. *Nature Geoscience, 3*, 756–761. doi:10.1038/ngeo979

Vigoa, M. I., Sanchez-Realesa, J. M., Trottini, M., & Chaoc, B. F. (2011). Mediterranean Sea level variations: Analysis of the satellite altimetric data, 1992–2008. *Journal of Geodynamics, 52*(2011), 271–278.

Chapter 3
Climate Change Impacts

Abstract According to the available data, climate-related changes have already been observed globally as well as in the eastern Mediterranean region. Projections about climate change are surrounded with a level of uncertainty regarding the pace and the magnitude of these changes. In any case, climate change is associated with a wide range of consequences, such as precipitation decrease, sea-level rise, ocean acidification, droughts, glaciers loss, melting of snow and ice sheets and extreme weather events such as heatwaves, floods and storms. These primary effects of global warming are leading to successive cumulative impacts on various sectors. Cyprus already experiences changes in the mean temperature and precipitation, and projections show exacerbation of the current situation. This chapter highlights the major impacts and focuses on the most vulnerable sectors for Cyprus: water resources, ecosystems and biodiversity, forests, agriculture, human health, coastal zones, energy supply and demand, tourism, and social impacts.

Keywords Agriculture · Climate impacts · Coastal zones · Ecosystems · Energy · Forests · Human health · Tourism · Water resources

According to the data presented in the previous chapter, climate is changing and climate-related changes have already been observed globally as well as in the eastern Mediterranean region. Projections about climate change are surrounded with a level of uncertainty regarding the pace and the magnitude of these changes. The uncertainty involves both the type and degree of the associated impacts of climate change. To this uncertainty (which is largely associated with the lack of information about the regional and local effects of climate change), one has to add eventual chain effects as well as the potential occurrence of new unforeseen effects due to accelerated glaciers retreat and large-scale disturbances in oceans, atmospheric circulation and pressure patterns.

As outlined in Chap. 2, climate change is associated with a wide range of consequences, such as precipitation decrease, sea-level rise, ocean acidification, droughts, glaciers loss, melting of snow and ice sheets and extreme weather events such as heatwaves, floods and storms. These primary effects of global warming are leading to successive cumulative impacts on various sectors.

Cyprus already experiences changes in the mean temperature and precipitation, and projections show exacerbation of the current situation. This chapter highlights the major impacts and focuses on the most vulnerable sectors for Cyprus:

- Water resources,
- Ecosystems and biodiversity,
- Forests,
- Agriculture,
- Human health,
- Coastal zones,
- Energy supply and demand,
- Tourism and
- Social impacts.

Table 3.1 provides a summary of the type and magnitude of these effects for each indicator, presented in the report 'Future Impacts of Climate Change across Europe' prepared for the Centre for European Policy Studies, in 2010. It is evident that Mediterranean Europe is expected to experience the most adverse climate change effects compared to other European regions. Although there have been more recent assessments of climate change impacts, such as those of the European Environment Agency (EEA 2012) and other studies mentioned in Sect. 2.2, Table 3.1 is concise and illustrative, and its findings have only been confirmed by subsequent reviews.

The next sections of this chapter provide an outline of the expected impacts for each sector mentioned above. All sections follow a similar pattern: starting from

Table 3.1 Simplified summary of climate change impacts in Europe and their intensity

Climate change indicators	Northern Europe	Central and Eastern Europe	Mediterranean
Direct losses from weather disasters	M(−)	M(−)	H(−)
River flood disasters	M(−)	H(−)	L(−)
Coastal flooding	H(−)	M(−)	H(−)
Public water supply and drinking water	L(−)	L(−)	H(−)
Crop yields in agriculture	H(+)	M(−)	H(−)
Crop yields in forestry	M(+)	L(−)	H(−)
Biodiversity	M(+)	M(−)	H(−)
Energy for heating and cooling	M(+)	L(+)	M(−)
Hydropower and cooling for thermal plants	M(+)	M(−)	H(−)
Tourism and recreation	M(+)	L(+)	M(−)
Health	L(−)	M(−)	H(−)

Source Behrens et al. (2010)
Notes H high, *M* medium, *L* low; (+) positive impact; (−) negative impact

findings of the 5th IPCC Assessment Report and other international and European studies, we gradually zoom in the Mediterranean region, first by providing evidence from studies in neighbouring countries such as Greece and Israel, and then describing the results—if any—from specialised studies carried out for Cyprus.

3.1 Water Resources

Temperature rise has already affected the water cycle globally, as observed by the occurrence of events such as:

- Reduced precipitation,
- Concentrated rainfall into extreme events,
- Increased evaporation and changes in evapotranspiration rates,
- Altered patterns of snowfall as well as melting of snow and ice caps,
- Increased frequency and intensity of droughts and
- Increasing water temperature and changes in soil humidity.

The above events pose direct and cumulative impacts on water resources. They lead to:

- water scarcity,
- increased water demand for agriculture,
- reductions in surface water and rivers run-off,
- reduced ground water recharge,
- fluctuated recharge of water reservoirs,
- increased risk of further salination of underground water,
- water quality deterioration,
- flash urban flooding,
- pollution and/or drying of wetlands and
- intensification of desertification.

Water resources under stress consecutively pose adverse effects on other sectors of the economy such as tourism, industry, agriculture and food production, and on the environment; they may additionally lead to further adverse societal impacts, changes in land use and economic activities and stronger urbanisation. The socio-economic impacts of droughts may arise from the interaction between natural conditions and human factors, such as changes in land use and land cover, water demand and use. Excessive water withdrawals can exacerbate the impact of drought (IPCC 2007a, 2014).

A review of Mediterranean and regional studies on climate change impacts points out that:

- The Mediterranean is very likely to face prolonged droughts and hence suffer from increasing water scarcity, declining crop yields and desertification (EEA 2012; IPCC 2014).

- The Mediterranean will face problems of water scarcity more than any other region in Europe. The decline in water availability is expected to be most pronounced in the Mediterranean and Southern Europe. Water availability may fall by 20–30 % under a 2 °C increase global warming scenario and by 40–50 % under the 4 °C scenario. An increase of 1–2 °C and a decrease of 10 % in precipitation, for instance, could lead to a decrease of 40–70 % in the annual average flow of rivers, which will impact agriculture, water and energy supply (Behrens et al. 2010, JRC 2014).
- Summer water flows may be reduced by up to 80 %, and the annual average water run-off will decrease in Central and Eastern Europe and in the Mediterranean by 0–23 % up to the 2020s and by 6–36 % up to the 2070s (Behrens et al. 2010).
- Increases in water demand for agriculture have occurred mainly in Mediterranean areas and this is projected to continue, thus increasing competition for water between sectors and uses (EEA 2012).
- The stress on water resources, combined with higher summer temperatures, will probably lead to a shift in the main seasons for the tourism sector (JRC 2009, 2014).
- Problems related to the quality of drinking water will be exacerbated all over the Europe and particularly at places in Mediterranean (Behrens et al. 2010).
- Serious floods, which may become more frequent, could cause severe damages to ecosystems and extended soil erosion, aggravate extinction of plant and animal species and habitat loses (Behrens et al. 2010).
- Aquifers may experience a further decline, owing to a decrease in rainfall and changes in enrichment capacity due to increased evaporation and water consumption by plants and humans. Coastal aquifers may also face water quality deterioration due to sea water penetration from the projected sea-level rise (European Commission 2007).
- Higher water temperature and variations in run-off are likely to produce adverse changes in water quality affecting human health, ecosystems and water use. Lowering of the water level in surface waters and aquifers will lead to elevated concentrations of pollutants due to lower dilution. More intense rainfall will lead to an increase of pesticides and fertilisers run-off, polluting surface and underground waters. Furthermore, higher water temperatures may enhance the release into the atmosphere from surface water bodies volatile and semi-volatile compounds (e.g. ammonia, mercury, dioxins and pesticides) (IPCC 2007a, 2014).
- Higher surface water temperatures will promote algal blooms and increase the bacteria and fungi content. This may lead to a bad odour and taste in chlorinated drinking water and the presence of toxins. Water-related diseases are expected to rise with increases in extreme rainfall and temperature (IPCC 2007b).

The cumulative cost of climate change for drinking water supply has been estimated in the frame of a Greek climate impact study for the decades 2041–2050 and 2091–2100, based on three SRES scenarios A1B, A2 and B2. The cost for 2041–2050 corresponds to 0.89 to 1.32 % of GDP. For the second decade under study, 2091–2100, the reduction of GDP starts for the best-case scenario A1B from 0.51 % of GDP and reaches, under the worst-case scenario A2, 1.84 % of GDP, i.e. €4.3 billion (Bank of Greece 2011).

In Cyprus, serious water scarcity is highlighted by the Water Exploitation Index, which is the highest in the EU, indicating severe stress on water resources and unsustainable water use (Eurostat 2015; Zoumides et al. 2014). This index expresses the available water resources compared to the amount of water used in a country. Moreover, Cyprus reported long-term total annual freshwater resources of the order of 300 Mm^3 (million cubic metres), which correspond to the lowest amount of annual resources per capita in the EU (410 m^3 per inhabitant) (European Commission 2007; Eurostat 2015). The country has already experienced severe droughts and water scarcity events, the most recent one in 2008, when authorities imported water from Greece to satisfy drinking water demand and imposed restrictions on water supply for both households and agriculture.

Another important indicator for assessing the quality of water resources management is groundwater exploitation. Cyprus is severely overstressing groundwater resources since it is exploiting groundwater beyond what has been set as the ecologically acceptable limit. Each year between 2004 and 2013, more than 100 % of the nationwide available groundwater for annual abstraction was extracted (Eurostat 2015). Through detailed modelling, Zoumides et al. (2013) found that, on average, total agricultural water use was 506 Mm^3/year. Groundwater exceeded by 45 % the abstraction rates recommended in the island's river basin management plan.

As shown in Fig. 3.1, Cyprus increasingly relies on desalination for covering drinking water needs. Desalination is an energy-intensive process, which leads to increased CO_2 emissions jeopardising the compliance of the country with

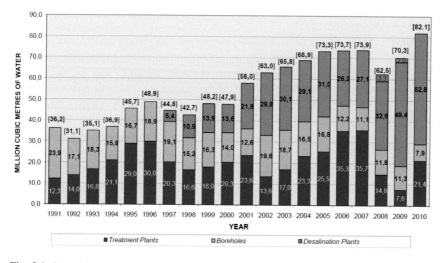

Fig. 3.1 Domestic water supply sources in Cyprus (1991–2010). *Source* Water Development Department (2011)

EU emissions reduction targets and any mitigation efforts (Lange 2011). Additional desalination plants have started operating after 2011, which led to a further dependence of the island's domestic water supply from this source.

Climate change in the region of Cyprus is expected to exacerbate the water-related challenges mentioned above in view of the expected decrease in rainfall and increase in temperatures, which will deteriorate aridity conditions. Apart from water availability, these conditions are likely to affect water quality as well, since the lack of adequate water resources may lead to overextraction of groundwater and hence to (a) further salinity of coastal aquifers and (b) high concentrations of pollutants in aquifers affected by industrial waste or the use of pesticides and fertilisers in agriculture.

According to a study that extends up to year 2030 only, annual costs of water shortages in the residential and industrial sector may reach 87.7 million Euros (at constant prices of year 2009), or 22 % higher than the costs without climate change (Zachariadis 2010a)—without taking into account the potential deterioration in water quality mentioned above.

Major water-related impacts of climate change are associated with the agricultural sector. These are addressed in more detail in Sect. 3.4.

3.2 Ecosystems and Biodiversity

Terrestrial (including freshwater) and marine ecosystems have already undergone changes due to climate change. The list of the impacts on ecosystems and biodiversity is wide and diverse forewarning for more adverse effects in the future. Projections and simulations on climate change, at the present state, do not take into account all factors influencing ecosystems and biodiversity, thus leading to variable and uncertain results.

Climate change impacts on ecosystems create chain effects that intensify global warming, since ecosystems play an important role in climate regulation by providing significant storage for carbon in peat lands, wetlands and the ocean and enhancing carbon sequestration from the atmosphere. Another example of climate change regulation via ecosystem services is that wetlands, salt marshes and dunes provide protection against flooding, coastal erosion and storms.

Apart from their ecological value, ecosystems provide various valuable services that human well-being depends on, such as food, water, energy, purification of water and air, primary production, cultural services etc. Overexploitation of these services and natural resources, inappropriate land use practices and unsustainable development have rendered ecosystems more vulnerable to climate change and thus less capable of adapting (European Commission 2010, 2011).

Main observed and projected impacts on ecosystems and biodiversity in Europe and the Mediterranean that are relevant for Cyprus, according to Behrens (2010), EEA (2008, 2012) and IPCC (2014) are the following:

- Accelerated changes since 2000 in phenology (timing of seasonal biological phenomena) and distribution of marine species have been observed, such as earlier seasonal cycles (by 4–6 weeks) and northward movements. Changes in geographic distribution of fish may affect the level of presence of commercially important fish species and the management of fisheries in general.
- Warming of surface water can have several effects on aquatic ecosystems like movement of freshwater species northwards and to higher altitudes and changes in life-cycle events (phenology). Observations indicate that spring phytoplankton and zooplankton bloom up to one month earlier than 30–40 years ago.
- Blue-green algae blooms, favoured by surface water warming, pose increased threats to the ecological status of lakes and reservoirs and enhanced health risks, particularly in water bodies used for public water supply and bathing.
- Climate change is responsible for the observed northward and uphill distribution shifts of many European plant species. Projection suggests that by late twenty-first century, plant species will shift several hundred kilometres to the north, forests are likely to have contracted in the south and expanded in the north and 60 % of mountain plant species may face extinction. The rate of change will exceed the ability of many species to adapt, especially as landscape fragmentation may restrict redistribution.
- Changes in the timing of seasons (e.g. the average advance of spring and summer between 1971 and 2000 was 2.5 days per decade) affect plant phenology. The pollen season starts on average 10 days earlier and is longer than what it used to be 50 years ago. Such changes are projected to continue in the future.
- Birds, insects, mammals and other animal groups are also moving to higher longitudes and altitudes. A combination of the rate of climate change, habitat fragmentation and other obstacles will impede the movement of many animal species, possibly leading to a progressive decline in European biodiversity. Distribution changes are projected to continue. Suitable climatic conditions for Europe's breeding birds are projected to shift nearly 550 km north-eastward by the end of the century, with the average range size shrinking by 20 %.
- Projections for 120 native European mammals suggest up to 9 % risk of extinction during the twenty-first century under a no-migration scenario.
- Advancement in the life cycles of many animal groups has occurred due to climate change, such as frog spawning, birds nesting and arrival of migrant birds and butterflies. These trends are projected to continue.
- Sea-level rise will reduce habitat availability for bird species that nest or forage in low-lying coastal areas.
- An assessment of European fauna indicates that the majority of amphibian (45–69 %) and reptile (61–89 %) species could expand their range under various SRES scenarios but since dispersion is very likely to be limited, it is more probable that the range of most species (>97 %) would contract.
- Plant species losses of up to 62 % are projected for the southern regions of Europe, particularly in the mountains where there may be a loss of endemism owing to invasive species as well as a loss in plant range.

- Mediterranean endemic plants and vertebrates are particularly vulnerable to climate change, especially the island species that have zero migration potential.
- The gradual rise in terrestrial and marine temperatures will cause the modification of natural habitats, which in the Mediterranean are already subject to intense pressures, and massive loss of biodiversity.

The Mediterranean Sea is undergoing tropicalisation, i.e. spread of thermophilic non-indigenous species (Lessepsian and Atlantic newcomers), and meridionalisation—a temperature-related change in indigenous species distribution. 745 species have been recorded in the Mediterranean Sea as valid aliens, so that the Mediterranean can be considered as one of the regions most severely affected by marine species invasions. Lessepsian migration through the Suez Canal affects not only the biota of the Levantine basin but, subsequently, the whole of Mediterranean. Lessepsian migrants represent over 12 % of the east Mediterranean's and 5 % of the entire Mediterranean's marine fauna (Zenetos et al. 2005).

Israel's Second National Communication on Climate Change concluded that there are plant species in Israel, which have adapted to stress conditions of heat and drought over the years, and can cope with a certain reduction in precipitation. Nonetheless, it is unclear whether these species will be able to survive an aggravation of conditions in the long term, especially an increase in dry periods, duration of heat loads, and increased frequency of drought years. Research studies have shown that a sharp decrease in precipitation in the south of Israel has already caused extensive changes in land cover and underlying biota, creating run-off and soil erosion. Species of blue-green algae have been recorded in the Sea of Galilee, due to the warming of surface water bodies. Climate change may also significantly impact on animal species. Higher temperatures and food shortages can contribute to a reduction in the body mass of birds and may have serious implications on bird populations' structure and species competition. Climate changes are amongst the factors that contribute to the spread and establishment of invasive tropical bird species in Israel (Israel's Ministry of Environmental Protection 2010).

According to the 4th National Report of the Republic of Cyprus to the United Nations Convention on Biological Diversity, the island has a rich biodiversity and is a centre of endemism for birds, mammals, insects and plants. The ecosystems in Cyprus include 48 habitat types, 14 of which are priority habitat types according to the Habitats Directive of the European Union (Directive 92/43/EEC) and 4 are endemic habitat types (CYPADAPT 2013). The Cyprus flora includes in total 1910 taxa native or naturalised, out of which 143 are endemic. The fauna contains 30 species of mammals, 25 species of amphibians and reptiles, about 375 bird species, 250 species of fish and about 6000 species of insects. The percentage of Cyprus endemism, calculating all taxonomical levels, is 7.39 %—one of the highest in Europe. 28.4 % of the area under the control of the Republic of Cyprus has been designated for inclusion in the "NATURA 2000" Network (Environment Department 2010).

The biodiversity of the island is affected by the invasion of alien species, epidemics from livestock-borne diseases affecting wildlife and from wildlife mismanagement practices. The number of threatened species at Cyprus level is 67 and 36 at international level (UES 2010). No data exist to clarify whether these species are threatened due to climate change.

An indicative case of the adverse effects on biodiversity caused by invasive species is the establishment of the Lessepsian migrant macroalga (seaweed) *Caulerpa racemosa* in many areas around the island, though it is not clear whether its expansion in the Mediterranean is attributed to climate change. It has spread extensively and replaced other species causing changes in the shallow water ecosystem flora and fauna (Argyrou et al. 1999; Demetropoulos 2002).

According to the vulnerability assessment conducted in the frame of the preparation of the National Adaptation Strategy of Cyprus, terrestrial ecosystems are estimated to be moderately vulnerable to climate change, whereas there are only few risks expected for marine and freshwater biodiversity. Terrestrial biodiversity seems to be threatened by landscape fragmentation of the island, as species cannot move neither northern nor higher after a certain point. Conversely, the main advantage of marine biodiversity is the ability of migration, which, however, can also be regarded as a disadvantage due to the intrusion of harmful invasive alien species. On the other hand, freshwater biodiversity is considered not to be threatened (CYPADAPT 2013).

3.3 Forests

Forests play a vital role in maintaining a stable global climate and environment. They purify water, influence rainfall patterns, protect aquifers and against extreme events, improve air quality and provide shelter to biodiversity. Forests offer numerous services and products with economic value. Moreover, they are a key factor in the carbon cycle and contribute to global warming mitigation acting as carbon sinks. Forest biomass in the EU contained 9800 million tonnes of carbon in 2010, an increase of 5.1 % compared to 2005 and of 26 % compared to 1990 (Eurostat 2011). Main forest species in Cyprus include the Turkish pine (*Pinus brutia*) and the black pine (*Pinus nigra*), which mainly cover the high-altitude areas of the Troodos Mountain. Endemic species include the golden oak (*Quercus alnifolia*) and the Cyprus cedar (*Cedrus brevifolia*), whereas there are numerous other native tree species on the island (Forestry Department 2011).

Forests in Europe will be affected by climate change, in terms of distribution, species composition, yields, storms and fires. In the south, forests will generally contract (EEA 2012). In particular, more frequent and severe summer droughts are likely to lead to reduced productivity, more extensive forest fires and, ultimately, desertification in some areas. Water scarcity and heat stress in the Mediterranean are expected to result in an increased frequency of forest mortality events, which will affect forest diversity (European Commission 2009a; Behrens et al. 2010).

Climate change is projected to cause substantial shifts in vegetation distribution. Changes in the distribution and the timing of seasonal events of both pests and pollinators will further affect forests, although the types of change are difficult to project. Periods of drought and warm winters are increasing pest populations and further weakening forests (EEA 2012).

Projected temperature increases will augment forest fire risk, leading to more ignitions and longer fire seasons. Predictions for forest fire risk under the SRES A2 and B2 scenarios analysed by the ENSEMBLES project for a 2 °C global temperature increase (2031–2060) show that an additional month of risk is expected over a great part of the Mediterranean, especially in inland locations (Van der Linden and Mitchell 2009). A reduction in precipitation or changes in its distribution will adversely impact forests causing plant to disappear, especially species which require more water, affecting forest diversity.

Important observations and projections about forests can be found in the study on climate change impacts in Greece report (Bank of Greece 2011). Three observed phenomena in Greek forests could be attributed to climate change or connected with it: Necrosis of fir trees; invasion of coniferous to broadleaf forests; and necrosis of *Pinus silvestris* (Scots pine) due to insects and fungus infestations, which are considered as secondary effects. As regards future projections:

- Coniferous and broadleaf evergreen forests will expand by 2–4 %, whilst the forests of spruce, fir, beech and black pine will shrink by 4–8 %, depending on SRES scenario (B2 or A2) for 2100.
- Due to reduction in forest productivity, a decline of wood biomass is expected, by 80,000 and 330,000 m^3 for B2 and A2, respectively.
- Timber production is expected to drop by 27 % (B2) to 35 % (A2) by 2100. The expected annual total reduction of wood biomass in Greece is 529,000–686,000 m^3, for scenarios B2 and A2, respectively, by 2100.
- Temperature increase will lead to more forest fires during summer time and the total burnt area will expand by 10–20 %, depending on the emissions scenario considered. As a result, the total cost of firefighting and forest restoration will increase by €40–€80 billion.
- It is projected that surface run-off and soil erosion will increase, leading to reduced water infiltration and consequently to poor enrichment of aquifers and reduction of the nationwide available water quantity by 5–8 billion m^3/year or 25–40 %.
- The present value of direct economic impacts of climate change on Greek forest ecosystems for the two SRES scenarios B2 and A2 ranges from €1.4 billion (B2) to €9.5 billion (A2).

Coming to Cyprus, forests suffered detrimental impacts during 2005–2008 due to drought, such as extended diebacks and necrosis and increased risk of fire ignitions (Forestry Department 2009). Although there are hardly any detailed simulations about the impacts of climate change on Cyprus forests, it is quite straightforward to foresee impacts due to temperature increase and precipitation decline. For example, climate simulations with the PRECIS model (Giannakopoulos et al. 2012) project a decline in rainfall during autumn in all areas of Cyprus where forests are located. The decrease in autumn rainfall can adversely affect forest growth because autumn follows a prolonged dry summer period in which forested areas are already under stress. The rise in summer temperatures and in the number of heatwaves mentioned in Sect. 2.4 will also have a negative impact on forest growth and will seriously increase the risk of forest fires. More specifically the following risks have been identified:

- Redistribution of species range due to contract and loss of habitats.
- Species extinction: *Pinus nigra* is expected to particularly suffer from climate change, owing to its very restricted distribution. Observations to date confirm these worries.
- *Pinus brutia* will be favoured from climate change and is expected to invade *Pinus nigra* habitats and displace the last.
- The national Forestry Department (2011) also expects species such as *Quercus infectoria* (a species of oak) and *Viburnum tinus* (a flowering plant) to be particularly threatened by forest fires and pests.

According to the National Adaptation Strategy of the Republic of Cyprus, the major vulnerability associated with forests (classified as 'moderate to high') relates to the dieback of specific tree species and the risk from insect attacks diseases—which have already occurred as explained above. The risk from forest fires is also considerable, whereas there seems to be no vulnerability from floods as forests in Cyprus are located in high-altitude areas only (CYPADAPT 2013).

3.4 Agriculture

Agricultural production is directly linked to climate; the type, quantity and quality of agricultural products depend heavily on it. Crop yields are mainly affected by rainfall patterns, rising temperatures, changing water availability and increasing carbon dioxide concentrations and the interaction amongst them. According to international and regional studies (Behrens et al. 2010; EEA 2008, 2012; FAO 2011; JRC 2014), several effects—largely caused by ongoing change in the climate—have already been observed:

- Expansion of forest fires already affects cultivation areas.
- Climate change affects the growing season, the timing of the cycle of agricultural crops (agrophenology) and average yields.
- Changes of the length of the growing season of several agricultural crops have been recorded, either lengthening or shortening depending on the latitude.
- The flowering and maturity of several species in Europe now occurs two or three weeks earlier than in the past with consequent higher risk of frost damage from delayed spring frosts.
- Many crops show positive responses to elevated carbon dioxide and low levels of warming, but higher levels of warming often negatively affect growth and yields.

- The variability of crop yields has increased due to extreme events, during the last decade, e.g. the heatwave of 2003 and the spring drought of 2007. Since extreme weather events are projected to increase in frequency and magnitude, crop yields will become more variable.
- Water demand for agriculture has increased by 50–70 % mainly in Mediterranean areas.
- Pest outbreaks, emergence of new pests and pathogens and increase in the frequency of diseases, as secondary effects, induced by higher temperatures, pose extra risks for crop production.

As regards future impacts, these studies lead to the following conclusions:

- Mediterranean countries will be severely affected by climate change and face most adverse effects on natural conditions for crop cultivation, leading to higher economic losses.
- The projected escalation in drought and heat stress risks in the period 2030–60 at 2 °C global warming is expected to have a negative effect on crop yields. More reduced yields are expected in the Mediterranean countries, which are most prone to a spread of drought risk. Limitations in irrigation water, due to drought, will further constrain crop production.
- More frequent occurrences of weather extremes, such as dry spells and heatwaves, will potentially damage agriculture more than changes in the annual average temperature.
- Under a scenario of 2.8 °C mean temperature increase in Europe, many areas of Southern Europe are projected to be less suitable for growing nearly all biofuel crops by 2050, and crops other than olives and heat and drought resistant ones.
- Temperature increase likely affects pollinating insect species leading to lower yields.

Giannakopoulos et al. (2009a) have found that increases in temperature and reduction in precipitation projected for both future climate scenarios (A2 and B2) over the Mediterranean basin in 2031–2060, when a 2 °C global warming is most likely to occur, lead to a substantial reduction of yields for all the crop types, through the reduction of the length of the growing period and the water available for crop growth. For rainfed crops, reductions in yields are more severe in the warmer southern Mediterranean than in the cooler northern Mediterranean, even when the fertilising effect of increased CO_2 was taken into account. The strongest impacts are generally observed for those crops growing during the summer period where the temperature increase is up to 4 °C and drought periods are longer. Furthermore, increased concentrations of tropospheric ozone lead to decreases in plant biomass and yields.

A recent study for the evaluation of the climate change impacts on Greek agriculture compared the projections based on SRES scenarios A1B, A2 and B2 for 2041–2050 and 2091–2100 with the reference period of 1991–2000, taking also into account potential effects from desertification (Bank of Greece 2011). Results revealed that:

- The B2 scenario is the most favourable for agricultural production;
- Wheat is the most sensitive arable crop;
- Cotton production will suffer the strongest yield reductions under both scenarios A1B and A2;
- The impact of climate change on tree crop yields by mid-century will vary from neutral to positive, worsening dramatically by the end of the century particularly for south Greece and the Aegean islands;
- Regarding pests, diseases and weeds, it is expected that higher temperature favours the development of pests, and the expansion of the occurrence of thermophilic weeds in colder zones and higher altitudes.

The agricultural sector of Cyprus is particularly vulnerable to climate conditions, due to its dependence on water resources. Unless adequate measures are taken for sustainable water management and without adaptation of agriculture to increasingly arid conditions, both rainfed and irrigated crops will be under higher stress in the future. Bruggeman et al. (2011) estimated that climate variability may lead to a 41–43 % reduction in total agricultural production for the period 2013/14–2019/2020, compared to 1980/81–2008/09—if at the same time irrigation water supply, a key factor for crop production, is reduced as a result of both climate change and water management policies aiming at sustainable use of water resources.

More recently, the AGWATER project, which was funded by national and EU sources, has carried out an in-depth analysis of the agricultural sector in view of climate change challenges. During that project, a digital soil map and a soil property database were created, along with an agrometeorological database. Then, a novel daily soil water balance model was developed and used to compute agricultural production and water use performance indicators in Cyprus for different economic/policy scenarios and three climate change scenarios consistent with the IPCC AR4 emission scenarios. According to the project results, irrigation water demand is expected to increase substantially by the mid-twenty-first century due to higher temperatures and increased rainfall. As a result, agricultural yields will decline and a reduction in net profits of crop production of up to 40 % is foreseen; economic losses are projected to be highest for rainfed crops (AGWATER 2014).

3.5 Human Health

Climate change is considered as a direct and indirect threat to public health. Although detailed regional assessments and projections on the type and extent of the impacts are surrounded by uncertainty, there is a clear agreement that harmful health impacts of climate change will be related to increasing heat stress, extreme weather events, poor air quality, undernutrition in less developed countries, and higher occurrence of water- and vector-borne diseases (IPCC 2014). The direct effects are caused by extreme weather events, and the indirect ones are a result of poor air and drinking water quality, diseases, food insecurity and ecological changes.

Europe has already faced increased mortality—during the 2003 heatwave. More than 70,000 additional deaths were reported in 12 European countries. Such heatwaves are projected to become more common later in the century as the climate continues to change, with mortality risk increases of between 0.2 and 5.5 % for every 1 °C increase in temperature above a location's specific threshold (EEA 2008). Moreover, the World Health Organization confirms that the distributional patterns of vector-borne infectious diseases (e.g. West Nile virus) are influenced by climate and are gradually expanding to northern latitudes (WHO 2003).

Main observations and projections are as follows:

- The World Health Organization on climate change impacts on Europeans' health notes that a 1 °C temperature increase is expected to result in increased mortality by 1–4 %. Mortality due to higher temperatures could increase by 30,000 deaths annually by 2030 and 50,000–110,000 deaths annually by 2080 (Bank of Greece 2011). This is confirmed by findings of the PESETA II project (JRC 2014).
- The rise in temperature and heatwaves will add significantly to heat-related deaths, in particular in Southern and South-eastern Europe and especially amongst elderly people. Considering the ageing of the population this might be a severe problem for the Mediterranean (Behrens et al. 2010; Bank of Greece 2011; JRC 2009, 2014).
- Potential rise in humidity, as a result of climate change, along with high temperature, contributes to heat stress and consequent human discomfort. Humidity has a greater effect on daily mortality and morbidity than temperature. The IPCC projects increases in heat-related mortalities from a baseline of 5.4–6 per 100,000 to a range of 19.5–248 per 100,000 by the 2080s (IPCC 2007b, 2014).
- The number of deaths attributable to cold weather could decrease (JRC 2009; Behrens et al. 2010).
- Deterioration of air quality in urban areas (due to forest fires, heat-induced increase in tropospheric ozone concentrations, increase in particulate concentrations) will lead to increased morbidity and mortality because of allergies and respiratory diseases (IPCC 2007b, 2014; JRC 2014; Bank of Greece 2011).
- Incidents of vector-borne diseases (West Nile fever, Leishmaniasis) spreading northward are expected to increase in the near future. Vector-borne diseases such as malaria or dengue fever could spread in European regions, and at higher altitudes. Additionally, some water- and food-borne disease outbreaks are expected to become more frequent with rising temperatures and more frequent extreme events. The risk of additional salmonella problems from bathing water quality is likely to grow (Behrens et al. 2010).
- Enhanced allergy risk is associated with rising temperatures and carbon dioxide concentrations, which lead to increased pollen production and prolonged pollen season in plant species with highly allergenic pollen (WHO 2003).

- Finally, one should not neglect the potential effects on human health because of increased levels of displacement and migration of human populations due to sea-level rise and increased frequency of weather extremes.

In Israel, as reported in its National Communication on Climate Change, increases in temperature, extreme precipitation events and heat stress in urban areas could favour a mosquito population increase and a change in their distribution, contributing to an outbreak of West Nile Fever (Israel's Ministry of Environmental Protection 2010).

Currently, there seem to be limited data or projections regarding climate change impacts on human health especially for Cyprus. National authorities and researchers did not possess clear relevant evidence up to the time of this writing. Based on internationally available information, the vulnerability assessment carried out in the frame of the National Adaptation Strategy of Cyprus comes to the same conclusion with that of the studies mentioned above. The major threat to public health comes from increased occurrence of heatwaves and higher temperatures. Other potential causes of concern, such as water- and food-related diseases and injuries from floods or landslides, are considered insignificant in that study (CYPADAPT 2013).

3.6 Coastal Zones

Coastal areas are experiencing adverse consequences related to climate and sea-level rise and are expected to be exposed to further risks over the coming decades. In short, anticipated climate-related impacts include:

- Inundation, flood and storm damage,
- larger extreme waves and storm surges,
- erosion,
- sea water intrusion and aquifer salination,
- sea water intrusion in estuaries and deltas and
- degradation or loss of coastal ecosystems and wetlands

The impact of climate change on coasts is exacerbated by increasing human-induced pressures. The direct impacts of human activities on the coastal zone have been more significant over the past century than impacts directly attributed to climate change. Major direct impacts include drainage of coastal wetlands, deforestation, sewage discharge, flow of fertilisers and contaminants into coastal waters, extractive activities, harvests of fisheries and other living resources, introduction of invasive species and construction of seawalls and other structures (dams) (IPCC 2007b, 2014).

Coastal flooding is expected to affect the Mediterranean considerably; mounting damages from flooding and extreme events such as storm surges and tsunamis are projected (IPCC 2014). It is projected that the expected rise in sea level will entail the submersion of low-lying coastal areas in the Mediterranean, causing various problems. Under a scenario of a 0.5 m rise in the sea level by 2100, the population exposed to the risk of coastal flooding will double in certain regions (Athens, Naples, Lisbon and Barcelona), leading to a considerable loss of assets. Foreseen erosion and flooding are expected to cause damage to coastal infrastructures (roadways, airports, sewerage systems, etc.), resulting in considerable economic losses (Behrens et al., 2010).

Assessments of the economic consequences of climate change (EEA 2008, 2012) state that by 2100, the population in the main coastal European cities exposed to sea-level rise and associated impacts on coastal systems is expected to be about 4 million and, if no adaptation measures are taken, assets exposed to sea-level rise may amount to more than 2 trillion Euros.

In the frame of the PESETA project, the number of people annually affected by sea floods was estimated for scenarios with global sea-level rise (SLR) from 48 to 58 cm by the end of the twenty-first century. An 88-cm SLR scenario has also been studied as a result of a higher emissions scenario. The number of people annually affected by sea floods in the reference year (1995) is estimated to be 36,000 and without adaptation increases significantly in all scenarios, in the range of 775,000–5.5 million people. The economic costs to people who might migrate due to land loss (through submergence and erosion) are also substantially increased under a high rate of sea-level rise, assuming no adaptation, and increase over time (JRC 2009).

Recent calculations for Greek coastal areas regarding the effects of SLR, both in the form of progressive SLR and wave tides, show that the impacts are expected to be particularly significant in the coming decades in Greece. According to a projection based on 0.5-m SLR by 2100, 15 % of the current total area of coastal wetlands in Greece is expected to be flooded. The estimated economic losses from erosion (for land uses: urban, tourist, wetland, forest and agricultural) for the entire Greek territory, for 2100 amounts approximately 356 million Euros for 0.5-m SLR and approximately 649 million Euros for 1-m SLR (Bank of Greece 2011).

Research studies have shown that sea level in the Israeli coast is expected to rise by some 0.5 m by 2050 and by approximately 1 m by 2100. An increase in the height and intensity of waves which penetrate inland, due to the increased intensity of extreme weather events, will increase the penetration of sea water in lower areas and will cause damages to water resources, coastal ecosystems and infrastructure. Projections suggest that a 10-cm rise will lead to a 2–10-m retreat of the coastline and to the loss of 0.4–2 km^2 of coast every 10 years. An increase in sea level of 1 m will flood a 50–100-m-wide belt on sandy beaches, which constitutes more than half the length of the Israeli coastline. One estimate, based on a scenario in which such an increase will take place until the year 2060, predicts that 8.4 km^2 of beaches will be lost, with an economic damage of NIS 4–5 billion (about 1 billion Euros). Additionally, changes in sand abrasion and accumulation process could shift the coastline inland (Israel's Ministry of Environmental Protection 2010).

Sea-level rise will worsen the sea penetration into aquifers, the majority of which already suffers due to overpumping. Coastal erosion could render coastal aquifers more vulnerable to sea water penetration. The Israeli government estimates that the loss of groundwater due to a possible sea-level rise of 50 cm could reach 16.3 Mm^3/km of coast. This loss will worsen if the frequency of droughts increases, during which the recharge rate of the aquifer will lessen. The expected rise in sea level and in the frequency and intensity of winter storms will cause severe damage to the coastal cliff of Israel, which will continue to retreat during the current century by tens of additional metres. Cliff retreat will cause extensive economic damage, including damage to existing properties and infrastructure (Israel's Ministry of Environmental Protection 2010).

The coastal zone of Cyprus, defined as 2-km inland from the coastline, represents 23 % of the country's total area. The island has a total shoreline of 735 km, of which about 385 km are under Turkish occupation since 1974. 10 km of coastal zone below 5 m elevation is less than 5 % of the Cypriot coastline, and the shoreline subject to erosion is 110 km—30 % of the coastline under control of the Republic of Cyprus (2006). The coastal zone is characterised by rich wildlife and marine and shore areas of high ecological and scientific value (Environment Department 2010).

There is considerable uncertainty about sea-level rise, since reports argue that vertical land movement is counteracting this potential effect. Erosion is considered to be a greater threat than flooding for Cyprus, the sandy and gravel beaches of the island being the most vulnerable ones. The coastline is already subject to erosion, as a result of human activities such as sand mining, dam and illegal breakwater construction and urbanisation. Climate change impacts could exacerbate coastal erosion (Republic of Cyprus 2006). No studies have been accomplished yet to clarify whether coastal erosion in Cyprus is also attributed to climate change. Increased erosion and sea-level rise could worsen the serious problem that Cyprus faces with sea water penetration to coastal aquifers and their salination.

The most vulnerable areas to possible sea-level rise, likely to face inundation risk and greater exposure to storm surges, are the low-lying area of Larnaca and the adjacent salt lake, the Akrotiri peninsula wetland, the Akamas Coastal/Marine Protected area and especially the Lara/Toxeftra Turtle Reserve, Cape Greko marine caves, and Poli Chrysochous coastline (Republic of Cyprus 2006; Parari 2009). It is important to note that most of these areas are protected under various EU Directives and international Conventions (i.e. "NATURA 2000", Ramsar). On the other hand, some of the most important infrastructures of Cyprus are located in low-lying coastal areas like the Larnaca airport, the desalination plant as well as the major power generating stations.

According to the National Adaptation Strategy of Cyprus, the main vulnerability of coastal zones in Cyprus to climate change is related to coastal erosion, which already constitutes an issue for the island as explained above. The impact of other climate change-induced effects such as coastal storm flooding and inundation is considered to be limited (CYPADAPT 2013).

3.7 Energy Supply and Demand

Climate change will have a direct effect on both the supply and demand of energy. Positive effects are expected for hydropower production in Northern Europe due to higher precipitation amounts and glacier melt. On the contrary, hydropower production in Southern Europe is projected to decrease by 25 % or more by 2050 and up to 50 % by the 2070s, as the hydropower sector highly depends on water. Decreased precipitation and increasing temperatures of the atmosphere and rivers are expected to influence negatively the cooling process of thermal power plants. Furthermore, extreme heatwaves can pose a serious threat to uninterrupted electricity supply, mainly because cooling air may be too warm and cooling water may be both scarce and too warm. As a result, energy demand may not be able to be met in the warm period, and additional capacity may need to be installed adding extra cost to energy production. The efficiency of photovoltaic plants could be slightly reduced by higher temperatures, especially during heatwaves. On the other hand, projected higher wind speeds could improve the electricity output of wind power plants. Additionally, extreme weather events, flooding and storm surges could damage infrastructure in vulnerable areas causing power outages (IPCC 2014; EEA 2008, 2012; Behrens et al. 2010).

On the demand side, increasing summer peaks for cooling and impacts from extreme weather events will affect in particular the electricity sector. Changes in demand will differ across Europe according to each region's climatic conditions. According to findings of the PESETA II project (JRC 2014), demand for heating in winter is expected to decrease in higher latitudes whilst demand for summer cooling will increase in Southern Europe and the Mediterranean. The net effect is projected be negative (i.e. an overall decline in energy demand), as increases in electricity demand for cooling are most likely to be outweighed by higher reductions in the need for heating energy. It is important to note that electricity required for cooling is generally more carbon-intensive than energy used for heating, the latter coming from non-electricity fuels in most European countries. Depending on the final energy mix of the electricity supply in Member States, net CO_2 emissions could even increase, thereby negatively affecting emission mitigation efforts.

There is considerable agreement amongst the existing EU-wide studies that Mediterranean countries will be faced with serious problems related to energy demand for summer cooling. Apart from the overall rise in cooling demand mentioned above, projections anticipate higher peaks in electricity demand in order to respond to the needs of summertime space cooling—by up to 50 % in Italy and Spain and 30 % in Greece by the 2080s (Behrens et al. 2010). Moreover, buildings in the Mediterranean region may need heating 2–3 fewer weeks per year; on the other hand, they will need 2–4 additional weeks for cooling by the mid-twenty-first century (Giannakopoulos et al. 2009b). Heat generated by air-conditioning units could raise ambient temperatures further and enhance the urban heat island phenomenon, thereby further increasing cooling demand in cities.

In Greece, winter time heating demand is forecast to be lower by 20 days/year for 2021–2050 and by 45 days/year for 2071–2100, whereas demand for cooling may rise by 10–20 extra days per year by 2021–2050 and 30–40 days by 2071–2100, compared to the reference period 1961–1990 (Bank of Greece 2011).

Coming to estimates for Cyprus, Lange (2011) focused on the interplay between growing demand for electricity and rising needs for desalinated water. Taking into account that electricity requirements for conventional sea water desalination are around 4.5 kWh/m^3 of desalinated water, and at a desalinated water production of 47.8 Mm^3/year, total electricity consumption needed for desalination amounted to 215 GWh/year around the year 2010. Keeping in mind that desalination capacity has grown since then and was expected in that study to reach 92.3 Mm^3 of desalinated sea water per year, electricity consumption was expected to rise to 415 GWh/year, which corresponds to about 10 % of total electricity consumption in recent years. Moreover, according to climate projections, 25 more days per year with temperatures above 35 °C will be added for the years 2021–2050 in comparison with those in the reference period 1961–1990, leading to an additional 25 days of cooling. Cooling and desalination needs lead to additional energy of about 670 GWh/year, which corresponds to more than 15 % of the country's annual electricity consumption. If other sectors of the economy, which might have increased energy demand owing to climate change impacts, are taken into consideration, the author estimated that the increase in energy demand could reach 20 % or even 30 % of the total electricity consumption in Cyprus for the period 2021–2050 relative to the 1961–1990 reference period.

Zachariadis (2010b) assessed the impact of climate change on electricity use in Cyprus up to 2030, based on the assumption that average temperature in the Eastern Mediterranean is expected to rise by about 1 °C by 2030. He found electricity consumption growing by 2.9 % in 2030 compared to a 'no climate change' scenario, with welfare losses (because of additional electricity needed to achieve the same thermal comfort with the 'no climate change' case) reaching 15 million Euros in 2020 and 45 million Euros in 2030 (at constant prices of year 2007). Additional peak electricity load requirements in the future because of climate change were estimated at 65–75 MW in 2020 and 85–95 MW in 2030, indicating increased requirements for reserve capacity.

In a more recent study, extending to year 2050 and combining an econometric model of electricity demand with detailed results of the PRECIS Regional Climate Model, Zachariadis and Hadjinicolaou (2014) found that annual electricity demand may rise by 6 % compared to a 'no climate change' case. This may cause annual welfare losses of up to more than 150 million Euros (at 2010 prices) because of increased energy expenditures of households and firms, and a total discounted cost between 488 and 732 million Euros'2010 for the entire period up to 2050. Although these additional power requirements are not very remarkable on an annual basis, the cost figures imply that the country may need to forgo one or two years of economic growth in order to cope with extra electricity needs due to climate change. Moreover, climate change is expected to exacerbate the already existing imbalance between (low) winter and (high) summer electricity demand.

Without electrical interconnections with any other country, Cyprus may have to address this imbalance with investments in peak load generating capacity that will be under-utilised during most of the year—thus leading to additional costs.

Climate change may also increase the risk of failures in the electricity transmission system due to higher temperatures, higher humidity and deposition of dust on insulators, thus leading to a higher need for cleaning insulators, which results in more frequent outage of generating units or transmission lines and decreases of the available power. During heatwaves, sea water (which is the cooling agent of power generating units in Cyprus) is warmer, resulting to insufficient cooling of the generating units leading to less efficient—and therefore more costly—power generation.

3.8 Tourism

Tourism is an important economic sector in Europe. Climate change has the potential to alter tourism patterns in the continent by inducing changes in destinations and seasonal demand structure. If unsustainable forms of tourism continue, they may exacerbate the negative effects of climate change. Some expected impacts according to EU-wide studies are the following:

- Traditional summer destinations like the Mediterranean are likely to suffer from a decline in tourism in the longer term because of worsening weather and climatic conditions (EEA 2008, 2012; Behrens et al. 2010).
- A climate change scenario of 1 °C increase could lead to a gradual shift of tourist destinations further north and up mountains, where outdoor activities would become more attractive, affecting summer tourism on the Mediterranean beaches (Behrens et al. 2010).
- Climate change will affect the tourism industry in the Mediterranean, which is of great economic importance for the region. Higher temperatures in 2031–2060 will particularly influence Mediterranean coastal zones, where a gradual decrease in summer tourism is projected, due to heatwaves and water supply problems. It is expected that the thermal comfort of tourists and their ability to acclimatise to a region prone to high temperatures and heatwaves may be heavily affected, resulting in a gradual decrease in summer tourism; at the same time, it could lead to an increase in spring and autumn tourism (Giannakopoulos et al. 2009a).
- Adjustment and redistribution or seasonal shift away from the summer peaks towards spring and autumn is a necessity; otherwise, the Mediterranean tourist sector will face decreasing bed nights (Behrens et al. 2010).
- South Europe, which currently accounts for more than half of the total EU capacity of tourist accommodation, would face a decline in bed nights, estimated to be in a range between 1 and 4 % by the 2080s compared with the 1970s, depending on the climate scenario, which is translated to losses ranging from 1.8 to 12.9 millions depending to the temperature scenario (JRC 2009).

- Water availability per capita decreases both in scenarios A2 and B2 by 2080s (JRC 2009) for coastal areas in the Mediterranean. If the current trend in tourist arrivals continues in the future, it is likely to have an increase of the pressure on water resources. The seasonality of the tourism and its spatial concentration could increase the stress on water availability, use and management. Water scarcity could in turn affect the sector even more than the consequences of climate change on tourist comfort, substantially limiting its growth and sustainability in some areas (JRC 2009). Temperate and tropical islands with high tourist arrival numbers and limited water resources are more likely to face water conflicts (OECD 2011).
- Tourism revenues across Europe, according to econometric analysis on the basis of two climate change scenarios examined in the PESETA II project, are expected to fall by 15 billion Euros per year, almost equally spread between South Europe (6 billion), northern Central Europe (4 billion) and South Central Europe (5 billion). Only the UK and Ireland are projected to experience a slight increase in revenues from tourism (JRC 2014). Barrios and Ibañez (2015) agree with these findings and estimate that the loss in revenues in Mediterranean Europe could lower their GDP by 0.45 %/year in the end of the twenty-first century.

The tourism sector may also face indirect impacts due to:

- Coastal flooding and erosion that will cause damages to infrastructure,
- Biodiversity loss and ecosystems deterioration,
- Deterioration of drinking water quality,
- Emergence of water-, food- and vector-borne diseases,
- Changes in air quality and
- Increased fire risk.

Tourism in Greece is expected to face negative effects due to climate change mainly during July and August as the mean air temperature and frequency, intensity and duration of heatwaves will increase. If the tourist period is extended to months May and September, these negative impacts can largely be offset, resulting in minor overall changes in the number of arrivals by the year 2100 (Bank of Greece 2011).

Cyprus will most likely face a decrease in summer tourism, associated with a seasonal shift in tourist distribution in spring and autumn, as it is projected for the Mediterranean region (Giannakopoulos et al. 2009a). Provided that proper adaptation measures are taken to make the island more attractive for off-peak season tourists, the economic impact may not be large. Furthermore, tourism-related water use represents 10–20 % of total domestic water use (Gössling 2006), and the stress of tourism to water resources available for irrigation may constitute the largest climate change-related vulnerability of the island with respect to tourism (CYPADAPT 2013).

3.9 Other Socio-economic Impacts

The social impacts of climate change are of particular importance, yet have not been studied in depth. Any policies and measures proposed must take into consideration their social implications if they are to be implemented effectively. Social injustice at national or global level will exacerbate the effects of climate change.

Chancel and Piketty (2015) demonstrate that inequality in carbon emissions exists both between countries and within countries: Top 10 % carbon emitters contribute to about 45 % of global emissions, whilst bottom 50 % emitters are responsible for 13 % of global emissions only. Top 10 % emitters live on all continents, with one-third of them from emerging countries. Despite this observation that is relevant for the *causes* of climate change, as regards the *social impacts* of climate change it is generally reasonable to distinguish two groups of countries: the developing and the developed world. The impacts are different in these two regions; in many places of the developing world, climate change-induced effects might be a matter of survival. For the poorest and most vulnerable people in today's world, the injustice of climate change impacts can still be summarised by the phrase: 'they didn't cause it, they are most affected by it, and they are least able to afford even simple measures that could help protect them from those damaging impacts that are already unavoidable' (Oxfam 2008).

Expected socio-economic impacts due to climate change in the developed world, particularly in the Mediterranean region, which have not been addressed in the previous sections, may be grouped under the following categories (Bank of Greece 2011):

- Reduced security due to the possible extreme weather events and fires,
- Changes in mortality and morbidity, impacts on labour productivity,
- Job losses in affected economic sectors (agriculture, fisheries and tourism),
- Reduced demand for recreational services due to changes in tourist distribution,
- Water shortage—declined standard of living because of water supply limitations,
- Loss of profit due to ecosystems damage,
- Capital losses due to infrastructure damages,
- Capital losses due to infrastructure and people relocation,
- Increased insurance rates,
- Migration from climate change affected countries of the developing world and civil unrest and
- Receiving and care of environmental refugees.

Limited water availability already poses problems in many parts of Europe and the situation is likely to deteriorate further due to climate change, with Europe's high water stress areas expected to increase from 19 % today to 35 % by the 2070s. This increase could accelerate migration pressures (European Commission 2009b).

References

AGWATER. (2014). Options for sustainable agricultural production and water use in Cyprus under global change—Project summary. Funded by the Republic of Cyprus and the European Regional Development Fund, Nicosia, Cyprus, September. Available at http://www.cyi.ac.cy/eewrc/eewrc-research-projects/climate-change-and-impact.html

Argyrou, M., Demetropoulos, A., & Hadjichristophorou, M. (1999). Expansion of the macroalga *Caulerpa racemosa* and changes in soft bottom macrofaunal assemblages in Moni Bay, Cyprus. *Oceanologica Acta, 22*, 517–528.

Bank of Greece. (2011). *Environmental, economic and social climate change impacts in Greece*, Athens, Greece. ISBN 978-960-7032-49-2.

Barrios, S., & Ibañez, J. N. (2015). Time is of the essence: adaptation of tourism demand to climate change in Europe. *Climatic Change, 132*, 645–660.

Behrens, A., Georgiev, A., & Carraro, M. (2010). Future impacts of climate change across Europe. Center for European Policy Studies (CEPS). Working Document No. 324, Brussels.

Bruggeman, A., Hadjinicolaou, P., Lange, M. A., Zoumides, C., Pashiardis, S., & Zachariadis, T. (2011). Effect of climate variability and climate change on crop production and water resources in Cyprus. The Cyprus Institute, Cyprus University of Technology, Cyprus Meteorological Service.

Chancel, L., & Piketty T., (2015. *Carbon and inequality: From Kyoto to Paris*. Paris School of Economics, Paris, France, November.

CYPADAPT. (2013). Report on the future climate change impact, vulnerability and adaptation assessment for the case of Cyprus. Deliverable 3.4, project CYPADAPT LIFE10 ENV/CY/000723. Available at: http://uest.ntua.gr/cypadapt/wp-content/uploads/DELIVERABLE3.4.pdf

Demetropoulos, A. (2002). Cyprus National Report on the Strategic Action Plan for the Conservation of Marine and Coastal Biological Diversity in the Mediterranean.

EEA (European Environment Agency). (2008). Impacts of Europe's changing climate—2008 indicator-based assessment, EEA Report No.4/2008, Copenhagen, Joint EEA-JRC-WHO report.

EEA (European Environment Agency). (2012). Climate change, impacts and vulnerability in Europe 2012. EEA Report No. 12/2012, Copenhagen, Denmark. ISBN: 978-92-9213-346-7, doi:10.2800/66071.

Environment Department. (2010). Fourth National Report to the United Nations Convention on Biological Diversity, Ministry of Agriculture, Natural Resources and Environment of the Republic of Cyprus, Nicosia.

European Commission, Directorate-General Environment. (2007). Water scarcity and droughts. In depth assessment; Second Interim Report.

European Commission. (2009a). Impact Assessment, Commission Staff Working Document accompanying the White Paper on Adapting to Climate Change: Towards a European Framework for Action (COM 2009, 147 final), SEC (2009) 387/2, Brussels.

European Commission. (2009b). *White Paper. Adapting to climate change: Towards a European framework for action*. COM(2009) 147 final.

European Commission. (2010). The European Union's Biodiversity Action Plan—Halting the loss of biodiversity by 2020—and beyond. Directorate-General Environment, Brussels, Belgium. ISBN 978-92-79-08071-5.

European Commission. (2011). COM(2011) 244 final. Our life insurance, our natural capital: an EU biodiversity strategy to 2020.

Eurostat. (2011). Forestry in the EU and the world. A statistical portrait. ISBN 978-92-79-19988-2.

Eurostat. (2015). Online information on the Water Exploitation Index, available at http://ec.europa.eu/eurostat/en/web/products-datasets/-/TSDNR310

FAO (United Nations Food and Agriculture Organisation). (2011). Potential effects of climate change on crop pollination, Rome, Italy.

Forestry Department. (2009). Short term action plan on combating the impacts of drought on state forests, Ministry of Agriculture, Natural Resources and Environment of the Republic of Cyprus, Nicosia.

Forestry department. (2011). Cyprus report: The State of the World's Forest Genetic Resources. Ministry of Agriculture, Natural Resources and Environment of the Republic of Cyprus, Nicosia.

Giannakopoulos, C., Le Sager, P., Bindi, M., Moriondo, M., Kostopoulou, E., & Goodess, C. M. (2009a). Climatic changes and associated impacts in the Mediterranean resulting from a 2 °C global warming. *Global and Planetary Change, 68,* 209–224.

Giannakopoulos, C., Hadjinicolaou, P., Zerefos, C., & Demosthenous, G. (2009b). Changing energy requirements in the Mediterranean under changing climatic conditions. *Energies,* *2*(4), 805–815. doi:10.3390/en20400805

Giannakopoulos, C., Petrakis, M., Kopania, T., Lemesios, G., & Roukounakis, N. (2012). Projection of climate change in Cyprus with the use of selected regional climate models. Deliverable 3.2 of CYPADAPT project funded by the EU Life Programme (project no. LIFE10 ENV/CY/000723). Available at http://cypadapt.uest.gr/?page_id=105

Gössling, S., Hall, C. M. (Eds.), (2006). *Tourism and global environmental change: Ecological, social, economic and political interrelationships.* London: Routledge.

IPCC. (2007a). Synthesis report. In Core Writing Team, R. K. Pachauri, & A. Reisinger (Eds.),Contribution of Working Groups I, II and III to the Fourth Assessment Report of the Intergovernmental Panel on Climate Change (104 pp.). Geneva, Switzerland: IPCC.

IPCC. (2007b). *Climate change 2007: Impacts, adaptation and vulnerability.* In M. L. Parry, O. F. Canziani, J. P. Palutikof, P. J. Van der Linden, & C. E. Hanson (Eds.), Contribution of Working Group II to the Fourth Assessment Report of the Intergovernmental Panel on Climate Change (pp. 315–356). Cambridge, UK: Cambridge University Press.

IPCC. (2014). *Climate change 2014: Synthesis report.* In Core Writing Team, R. K. Pachauri, & L. A. Meyer (Eds.), Contribution of Working Groups I, II and III to the Fifth Assessment Report of the Intergovernmental Panel on Climate Change (151 pp.). Geneva, Switzerland: IPCC. ISBN 978-92-9169-143-2.

Israel's Ministry of Environmental Protection. (2010). Israel's Second National Communication on Climate Change. Submitted under the United Nations Framework Convention on Climate Change.

JRC (European Commission's Joint Research Centre). (2009). Climate change impacts in Europe—Final report of the PESETA research project. Projection of Economic Impacts of Climate Change in Sectors of Europe based on Bottom-up Analyses. Available at http://ipts.jrc.ec.europa.eu/publications/pub.cfm?id=2879

JRC (European Commission's Joint Research Centre). (2014). Climate Impacts in Europe. The JRC PESETA II Project. JRC Scientific and Policy Reports, EUR 26586EN. Available at http://ipts.jrc.ec.europa.eu/publications/pub.cfm?id=7181

Lange M. A. (2011). The energy-water nexus on Cyprus. Energy, Environment and Water Research Center, The Cyprus Institute. Workshop: Energy Issues Facing Cyprus, Nicosia.

OECD/UNEP (Organisation for Economic Cooperation and Development & United Nations Environment Programme). (2011). *Climate change and tourism policy in OECD Countries,* OECD Studies on Tourism, Paris, France. doi:10.1787/9789264119598-en

Oxfam. (2008). Briefing paper: Climate, poverty, and justice.

Parari, M. (2009). Climate change impacts on coastal and marine habitats of Cyprus. Cyprus International Institute for the Environment and Public Health in Association with Harvard School of Public Health.

Republic of Cyprus. (2006). Report under the Chapter VI. 1-2 of the Recommendation 2002/413/EC concerning the implementation of Integrated Coastal Zone Management in Europe, 2006. A strategic approach to the management of the Cyprus coastal zone.

UES (Unit of Environmental Studies). (2010). Review of biodiversity research results from Cyprus that directly contribute to the sustainable use of biodiversity in Europe. Intercollege Research Centre, Nicosia, Cyprus.

Van der Linden, P., & Mitchell, J. F. B. (Eds.). (2009). ENSEMBLES: Climate change and its impacts: Summary of research and results from the ENSEMBLES project (160 pp). Met Office Hadley Centre, FitzRoy Road, Exeter EX1 3 PB, UK.

Water Development Department. (2011). Cyprus River Basin Management Plan (Annex VII), Ministry of Agriculture, Natural Resources and the Environment of the Republic of Cyprus, Nicosia.

WHO (World Health Organisation). (2003). Health and Global Environmental Change. Methods of assessing human health vulnerability and public health adaptation to climate change, Geneva, Switzerland. ISBN 92 890 1090 8.

Zachariadis, T. (2010a). Forecast of electricity consumption in Cyprus up to the year 2030: The potential impact of climate change. *Energy Policy, 38*(2010), 744–750.

Zachariadis, T. (2010b). Residential water scarcity in Cyprus: Impact of climate change and policy options. *Water, 2010*(2), 788–814.

Zachariadis, T., & Hadjinicolaou, P. (2014). The effect of climate change on electricity needs—A case study from mediterranean Europe. *Energy, 76*, 899–910.

Zenetos, A. et al., (2005). Annotated list of marine alien species in the Mediterranean with records of the worst invasive species. *Mediterranean Marine Science, 6*(2), 63–118.

Zoumides, C., Bruggeman, A., Zachariadis, T., & Pashiardis, S. (2013). Quantifying the poorly known role of groundwater in agriculture: The case of Cyprus. *Water Resources Management, 27*, 2501–2514.

Zoumides, C., Bruggeman, A., Hadjikakou, M., & Zachariadis, T. (2014). Policy-relevant indicators for semi-arid nations: The water footprint of crop production and supply utilization of Cyprus. *Ecological Indicators, 43*, 205–214.

Chapter 4
Adapting to Climate Change

Abstract Both mitigation and adaptation to climate change are required in order to strengthen climate resilience. Although Cyprus contributes a negligible amount of emissions to global warming, it is likely to suffer considerable climate change impacts because of its location in one of the most vulnerable regions of the planet. Therefore, the formulation and implementation of a strategy on adaptation to climate change are crucial for the country. This chapter outlines general principles for good climate change governance, in line with the European best practices; provides an overview of a realistic climate adaptation strategy; highlights the economic and fiscal aspects of adaptation and emphasises the importance of a green tax reform as a valuable measure to achieve both economic and climate resilience.

Keywords Adaptation economics · Adaptation strategy · Climate resilience · Fiscal policy · Green taxes

4.1 General Principles for Good Climate Change Governance

Adaptation to climate change refers to an adjustment in natural or human systems in response to actual or expected climatic stimuli or their effects, which moderates harm or exploits beneficial opportunities (IPCC 2007). Adaptation has become increasingly important in recent years and has gained political significance in international climate negotiations, especially because developing countries emphasise the linkage between adaptation and economic development. Integration of adaptation into planning and the design of policies can promote synergies with the development and reduce the risk of disasters. It is also clear that climate change mitigation and adaptation should be addressed jointly to ensure future climate resilience (IPCC 2014).

The EU Strategy on Adaptation to Climate Change (European Commission 2013a) identifies three priority areas: promoting action by Member States, through

© The Author(s) 2016
T. Zachariadis, *Climate Change in Cyprus*,
SpringerBriefs in Environmental Science, DOI 10.1007/978-3-319-29688-3_4

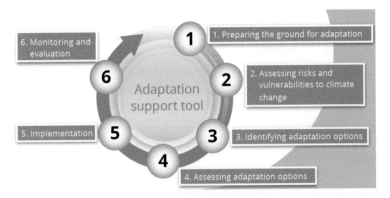

Fig. 4.1   The European adaptation support tool. *Source* EEA (2015)

encouraging the adoption of comprehensive adaptation strategies and providing funding to help them build up their adaptation capacities; 'climate-proofing' action by promoting adaptation in key vulnerable sectors and ensuring that Europe's infrastructure becomes more resilient and better informed decision-making by addressing gaps in knowledge about adaptation. Since the early 2000s, and following some good practices by European countries, the concept of how to formulate an effective adaptation strategy has evolved. The basic principles are summarised in a 'European adaptation support tool', which is illustrated graphically in Fig. 4.1 and formed the basis for EU-wide guidance on national adaptation strategies (European Commission 2013b). Several EU Member States have already adopted a National Climate Change Adaptation Strategy in line with these principles.[1]

National adaptation strategies have to be developed within a situation of changing socio-demographic trends (e.g. ageing population), changing technologies and lifestyles (e.g. increased dependence on transport modes and electricity-consuming communication technologies) and evolving forms of national and EU-wide governance PBL (2015). In particular, the governance issue, which is often considered as a non-technical and hence 'soft' topic, should not be underestimated. Climate change adaptation is widely recognised as a multi-level endeavour that requires the coordination of different levels of government (Bauer and Steurer 2014). According to the OECD, it is crucial to promote the governance of climate change across all levels of government and stakeholders, so as to avoid policy gaps between local action plans and national policy frameworks and to encourage cross-scale learning between relevant departments in local and regional governments. In a proper framework, national authorities should take the lead in designing proper climate action plans, build nationwide capacity and allow experimentation with individual adaptation measures at a local scale.

---

[1]The National Adaptation Strategy of Cyprus had not been formally adopted by the time of this writing (December 2015).

The implementation of specific adaptation measures may then be left out to local authorities, which are in a better position to engage citizens in strengthening climate resilience (Corfee-Morlot et al. 2009).

4.2 Adaptation Measures for Cyprus

Both mitigation and adaptation to climate change are required in order to strengthen climate resilience. It is widely accepted that even if greenhouse gas emission reduction targets are reached through global action, climate change will take place to some extent, because the emitted greenhouse gases will continue to trap heat and increase the earth's average temperature. Although Cyprus contributes a negligible amount of emissions to global warming, it is likely to suffer considerable climate change impacts because of its location in one of the most vulnerable regions of the planet. Therefore, the formulation and implementation of a strategy on adaptation to climate change are *sine qua non* for the country.

In general, the adaptation strategy has a preventive role, and the policies and measures aiming to tackle the impacts of climate change should be developed prior to the onset of those impacts. In order to reduce the possible harm to people, property, services and nature/ecosystems, a country must develop a preventive adaptation strategy, which should be revised and updated on a regular basis.

Financial constraints are considered as an inhibiting factor for the implementation of an adaptation strategy. In order to reduce the dependence of the implementation on any financial needs, it is advisable, at this stage, to set quick start actions, which are easy to implement and do not require any investments such as the horizontal integration of adaptation in all policies, full use of existing EU legislation and use of plans and projects already included in the state budget.

For an effective implementation, the adaptation strategy should define the authorities and stakeholders involved in the implementation, adopt a time schedule, make use of proper financing instruments and set a monitoring mechanism. As the adaptation measures will incur both private and public costs, a prerequisite for effective implementation of a strategy for adaptation is the collaboration between the public and private sector as well as public consensus and cooperation. As adaptation measures can only be region- or location-specific, adapting to climate change will involve all levels of participation and action—individual, social, regional and national level.

National and local authorities should regularly obtain and exchange information with other European regions in order to learn mutually from each other's adaptation efforts. The online European Climate Adaptation Platform (Climate-ADAPT) constitutes a valuable source of relevant information.[2]

---

[2]See http://climate-adapt.eea.europa.eu/.

The adaptation measures outlined below reflect the observed and projected impacts of climate change in Cyprus as summarised in the previous chapters.

- For *water resources*, the measures should enhance ecosystem storage capacity, protect surface and groundwater quality, promote good condition of soil, enhance water management and efficient water use and implement appropriate water pricing to reflect scarcity and environmental costs.
- For *biodiversity*, an inventory of all species has to be compiled, and especially those species that are sensitive to climate change should be monitored. Alien species must be recorded in order to prevent their expansion to the possible extent. Sustainable use of ecosystem services and natural resources should be promoted for this purpose.
- Special attention must be given to improve *forest resilience* to fires by classifying forests according to the risk of fire and establishing early warning systems and water supply systems for firefighting.
- For *agriculture*, sustainable use of water resources with special attention to halt overexploitation of groundwater and to improve irrigation management should be implemented. Sound use of chemical pesticides/herbicides and fertilizers must be applied to protect soil and water resources from pollution/degradation.
- For the *public health* sector, data must be collected, and an inventory must be completed on water and food-borne diseases, and disease control and prevention must be promoted. Health and social care systems must develop contingency plans to cope with potential disease outbreaks. The implementation of measures for air quality improvement in urban areas must be enhanced. Air and drinking water quality must be monitored, and strict inspections in food production and service industry must be applied.
- For *coastal zones*, research on sea-level rise must be promoted, in order to monitor vulnerable sites and simulate the eventual future effects. An inventory of coastal areas prone to erosion risk must be prepared. The possibility to implement additional erosion protecting measures should be examined, such as the protection of wetlands and sand dunes.
- For *energy supply and demand*, energy efficiency improvements are very important. Implementation of proper carbon pricing for all fuels (to be explained in Sect. 4.4) and raising of public awareness play a key role for encouraging energy conservation. Long-term electricity generation plans have to be adjusted in order to account for the additional capacity needed due to climate change, giving priority to renewable electricity generation.
- For the *tourism* sector, guidelines for adaptation to climate change must be developed. The tourism industry should rapidly respond to the expected decrease in summer tourism and to the shift in tourists' distribution to spring and autumn and must take action to combat the emerging competitiveness from other destinations in Europe, which will be favoured by climate change. Additionally, the tourism industry should make investments in infrastructure/technologies to upgrade facilities to face the increased temperature and water shortage.

- Climate resilience should also be an important ingredient of *spatial planning* policies. Authorities should encourage a gradual shift to low-carbon cities, investing in substantial energy refurbishment of the building stock (which is both a climate change mitigation and an adaptation measure) and using nature-based solutions (e.g. green areas and water harvesting) in order to reduce the vulnerability of the urban population to a hotter and drier climate.
- Society must be informed and mobilised in order to get prepared for climate change adaptation measures. Special attention should be paid to *socio-demographic trends* such as urbanisation, which could accelerate the spread of public health problems, and ageing, which will lead to an increased fraction of the population being vulnerable to heat stress and diseases. Funds should be foreseen to enable low-income households to afford capital expenditures for adaptation/mitigation measures, in order to avoid 'energy and climate poverty'. The country's Social Security Services need to prepare a proper strategy, and the insurance sector should consider developing new insurance products to cover climate change-related risks.
- Society should also be well informed about issues related to climate resilience. Proper *awareness-raising activities* will therefore include information regarding the following: impacts of climate change, measures for households and firms to adapt to these adverse impacts and co-benefits of adaptation action such as improvement in air quality and energy security and reduced expenditures for energy and water use.

A long (but still not exhaustive) list of climate change adaptation measures for Cyprus is provided in the Appendix.

4.3 Economic and Fiscal Aspects of Adaptation

As shown in Chap. 3, some of the eventual climate change impacts are known with more certainty than others. Pressure on water resources, energy supply, agriculture, forestry and tourism is well recognised in Cyprus, and national strategies to tackle these problems already exist to some extent. Examples of such policy responses are the national strategy for sustainable development (Environment Department 2010), the national strategy to combat desertification (Environment Department 2008), the sustainable water management strategy, legislation and grant schemes for improving energy efficiency in buildings and forest protection measures. In general, these policies are well designed and are in line with (or a consequence of) policy initiatives in the European Union. Successful implementation of these policies is a key priority in these sectors.

Other impacts, however, are associated with more uncertainty. Even if it seems likely that public health, biodiversity and coastal infrastructure will be threatened as a result of climate change, the extent and degree of such adverse effects are unknown, mainly because the monitoring and data collection infrastructure are inadequate or simply unavailable in the country. In these sectors, it is important to establish the monitoring mechanisms in order to provide much needed data which can send early warnings to policy-makers and the public and can help avoid large natural and economic damages. The associated economic analysis has to take account of these uncertainties as well. According to the IPCC (2014), economic thinking on adaptation has evolved from a focus on cost–benefit analysis and deterministic optimisation techniques to the development of multi-metric evaluations including the risk and uncertainty dimensions in order to provide meaningful policy support to decision-makers. Still, uncertainty and risk analysis are at an early stage of development: as Watkiss et al. (2015) remark, decision support tools available for the economic appraisal of adaptation measures are still based on deterministic approaches.

In times of fiscal consolidation, however, it is questionable whether governmental mechanisms can provide sufficient funds to implement the long list of adaptation measures presented in Sect. 4.2. In order to prioritise some actions, some remarks are necessary to keep in mind (Jones et al. 2012).

It is widely accepted that the cost of some adaptation measures can help avoid much larger costs in the future. To provide a few examples, changing agricultural production towards more heat-resistant crops, improving the energy efficiency of buildings and investing in order to make the country more attractive for winter tourism can prepare the economy for the future and thereby can reduce the costs of the impacts of climate change in the coming decades.

Apart from public sector costs, many adaptation measures will require expenditures from the private sector such as farmers, home owners and enterprises. Therefore, the direct fiscal cost of such measures is expected to be moderate— unless sudden and strong changes in the climate happen in the future which cannot be avoided through private investments only.

Still, private adaptation measures may prove to be inadequate due to insufficient information or unavailability of private funds, particularly if climate change-induced events are abrupt or irreversible (e.g. prolonged heatwaves, storms or floods). In this case, government intervention and increased public spending will be necessary in order to alleviate the damages. This underlines the need for authorities to install proper monitoring mechanisms and early warning systems in order to inform citizens (e.g. home owners and enterprises in coastal areas) about future risks. Moreover, the need for large public expenditures on infrastructure requires prudent management of this spending, probably within a medium-term financial framework consistent with the available resources, macroeconomic stability and debt sustainability (IMF 2015).

In order to make access to capital easier for the private sector, the government may consider providing economic incentives for climate change adaptation investments in vulnerable sectors (in the form of direct grants, aid or guarantees for specialised private insurance schemes).

4.4 Green Tax Reform: An Effective Environmental and Economic Adaptation Strategy

As the IPCC (2014) highlights, not all adaptation measures are investment related. Apart from the environmental regulations, specific policies promoting economic instruments have a high potential and can be less costly to the public budget. Such instruments are currently not well explored in an adaptation context. Amongst several economic instruments, the importance of pricing policies for climate change adaptation should be emphasised. This section provides more details about pricing instruments that can be very beneficial both for adaptation policies and for the economy as a whole.

Proper pricing of water and energy as well as charges for unsustainable use of resources (such as congestion charging) can become a key priority for mitigating climate change impacts. Pricing measures can reconcile economic growth and job creation with environmental protection, with the aid of a 'Green (environmentally friendly) Tax Reform' (GTR). By definition, a GTR (also called green fiscal reform) involves a reform of the national tax system whereby there is a shift of the burden of taxes towards environmentally damaging activities such as an unsustainable resource use or pollution.

A possible GTR for Cyprus may comprise introducing a carbon tax to apply on fuels used across all economic sectors and increasing taxation on the use of resources (e.g. water) and other environmentally harmful activities (e.g. waste production, air pollution and use of fertilizers and pesticides), whilst at the same time decreasing labour taxation, by reducing social security contributions of both employers and employees. As Cyprus has running balanced public budgets during the 2014–2015 period, this reform can be designed so as to prove revenue neutral: the extra revenues to be generated through environmental taxes can be roughly equal to the public revenues lost through the reduction in labour taxes. The Cypriot economy is labour intensive; hence, the reduction in labour taxation is expected to lower costs of most enterprises even despite the increase in energy costs; a few exceptions may need to be addressed in a targeted manner. Part of the additional revenues can also be used to compensate low-income households for the increase in their energy bills or finance green investments such as energy refurbishments in buildings or improvements in public transport infrastructure.

This reform can be implemented gradually over a longer period (e.g. five years) in order to provide time to firms and consumers to adapt to the new situation. Thanks to the quite low international crude oil prices during 2014–2015, a carbon tax will not increase fuel prices substantially, so that rising energy costs should be manageable by all households and firms.

As illustrated in Fig. 4.2, a Green Tax Reform has multiple environmental benefits. Individuals and enterprises will gradually adjust their investment decisions and consumption behaviour in order to adapt to the new tax system, thereby reducing the use of energy in industry, buildings and motor vehicles, and substituting

Fig. 4.2 The multiple benefits of a Green Tax Reform

towards low-carbon or zero-carbon energy sources. This in turn will do the following:

- Reduce the energy import dependency of Cyprus and thereby improve its persistent current accounts deficit (European Commission 2012);
- Improve air quality by reducing the emissions of other air pollutants too and
- Contribute to climate change adaptation since, e.g., a better insulated building is less vulnerable to high external temperatures, and a less water-intensive agricultural sector is less dependent on water availability.

More than 20 years of experience in a number of EU Member States show that a Green Tax Reform can be beneficial for the economy and employment. In several case studies around Europe, it has been clearly demonstrated that energy/environmental taxes are less detrimental to employment and growth than other direct and indirect taxes (European Commission 2013c; Vivid Economics 2012). If the environmental tax increases are accompanied by reductions in labour taxation, the overall effect can be beneficial in both macroeconomic and environmental terms. Such a reform reduces distortions on economic activity and changes relative economic prices, thereby fostering innovation and encouraging investment in green economy sectors, which can create a competitive advantage for the economy.

Nine countries (Belgium, Denmark, Finland, France, Germany, the Netherlands, Portugal, Sweden and the United Kingdom) had adopted such reforms until late 2015, whilst Ireland has employed partial measures in the same direction (carbon tax, plastic bag levy).

Major international organisations such as the OECD, the IMF and the World Bank have underlined that Green Tax Reforms are the most efficient way to ensure both fiscal sustainability and environmental protection (Hagemann 2012; Parry et al. 2014). This is also in line with the EU's strategic initiative for a resource-efficient Europe achieving a 'circular economy' (European Commission 2014). Policy-makers in Cyprus should seriously consider including such charges in future policy measures.

References

Bauer, A., & Steurer, R. (2014). Multi-level governance of climate change adaptation through regional partnerships in Canada and England. *Geoforum, 51*, 121–129.

Corfee-Morlot, J., Kamal-Chaoui, L., Donovan, M. G., Cochran, I., Robert, A., & Teasdale, P. J. (2009). Cities, Climate Change and Multilevel Governance. OECD environmental working papers No. 14, Organisation for Economic Cooperation and Development, Paris, France.

Environment Department. (2008). National Action Plan to Combat Desertification, Ministry of Agriculture, Natural Resources and Environment, Nicosia.

Environment Department. (2010). National Strategy on Sustainable Development, Ministry of Agriculture, Natural resources and Environment, Nicosia.

European Commission. (2012). Macroeconomic imbalances—Cyprus. occasional papers No. 101 of 'European Economy' series, Directorate-General for Economic and Financial Affairs. ISSN 1725-3209.

European Commission. (2013a). An EU Strategy on adaptation to climate change. Document COM(2013) 216 final, Brussels, Belgium. Available at: http://ec.europa.eu/clima/news/articles/news_2013041601_en.htm

European Commission. (2013b). Guidelines on developing adaptation strategies. Staff Working Document (2013) 134 final, Brussels, Belgium. Available at: http://ec.europa.eu/clima/policies/adaptation/what/docs/swd_2013_134_en.pdf

European Commission. (2013c). Tax Reforms in EU Member States 2013: Tax policy challenges for economic growth and fiscal sustainability, Report No. 5/2013 of 'European Economy' series, Directorates-General for Economic and Financial Affairs and for Taxation and Customs Union. doi:10.2765/40142

European Commission. (2014). Communication from the Commission to the European Parliament, the Council, the European Economic and Social Committee and the Committee of the Regions: Towards a circular economy: A zero waste programme for Europe, COM(2014)398 of 2.7.2014.

EEA (European Environment Agency). (2015). National monitoring, reporting and evaluation of climate change adaptation in Europe. EEA Technical Report No. 20/2015, Copenhagen, Denmark. ISBN ISBN 978-92-9213-705-2, doi:10.2800/629559.

Hagemann, R. (2012). Fiscal consolidation: Part 6. What Are the Best Policy Instruments for Fiscal Consolidation? OECD Economics Department working papers, No. 937, Paris, France.

IMF (International Monetary Fund). (2015). The Managing Director's statement on the role of the Fund in addressing climate change, Washington, DC, November. Available at http://www.imf.org/external/np/pp/eng/2015/112515.pdf

IPCC. (2007). Synthesis report. In Core Writing Team, R. K. Pachauri, & A. Reisinger (Eds.), Contribution of Working Groups I, II and III to the Fourth Assessment Report of the Intergovernmental Panel on Climate Change (104 pp.). *IPCC, Geneva, Switzerland.*

IPCC. (2014). Climate change 2014: Synthesis report. In Core Writing Team, R. K. Pachauri, & L. A. Meyer (Eds.), Contribution of Working Groups I, II and III to the Fifth Assessment Report of the Intergovernmental Panel on Climate Change (151 pp.). Geneva, Switzerland: IPCC. ISBN 978-92-9169-143-2

Jones, B., Keen, M., & Strand, J. (2012). Fiscal Implications of Climate Change. Policy Research Working Paper No. 5956, The World Bank Development Research Group, Washington, DC, January 2012.

Parry, I. W. H., Heine, D., Lis, E., & Li, S. (2014). *Getting energy prices right: From principle to practice*. Washington, DC: International Monetary Fund. ISBN 978-1-48438-857-0.

PBL (Netherlands Environmental Assessment Agency). 2015. *Adaptation to climate change in the Netherlands—Studying related risks and opportunities*. PBL publication number: 1632. The Netherlands: Hague.

Vivid Economics. (2012). Carbon taxation and fiscal consolidation: The potential of carbon pricing to reduce Europe's fiscal deficits. Report prepared for the European Climate Foundation and Green Budget Europe, May 2012.

Watkiss, P., Hunt, A., Blyth, W., & Dyszynski, J. (2015). The use of new economic decision support tools for adaptation assessment: A review of methods and applications, towards guidance on applicability. *Climatic Change,*. doi:10.1007/s10584-014-1250-9.

Chapter 5
Concluding Remarks

Abstract Cyprus is located in a hot spot as regards climate change and is projected to face significant temperature increases and decline in rainfall levels. As a result, the country is expected to be faced with serious challenges in agriculture, water resources, biodiversity, forests, energy demand, coastal zones and human health. The extent to which such effects will affect the society and economy of Cyprus depends crucially on policies to be implemented in the frame of a national adaptation strategy. Mitigation and adaptation policies should be closely linked and can offer synergies for coping with climate change. For this purpose, proactive actions need to be taken by both the public and the private sectors. Adequate monitoring mechanisms should be set up in order to provide much needed data which can send early warnings to policy makers and the public and can help avoid large natural and economic damages at a later stage. Enabling private adaptation investments and properly pricing the use of natural resources in the frame of a well-designed environmental fiscal reform are key priorities for investing in a climate resilient economy.

Keywords Climate adaptation · Climate monitoring · Climate resilience

According to the current scientific consensus, warming of the global climate system due to human activities is unambiguous. The Mediterranean basin is considered amongst the geographic areas that are most vulnerable to climate change and is expected to experience adverse climate change effects. Therefore, Cyprus is located in a hot spot and is projected to face significant temperature increases and decline in rainfall levels.

Forecasts from regional climate model simulations for the twenty-first century highlight the vulnerability of Cyprus to climate change by projecting an increase in maximum temperature of 1.3–1.9 °C for 2021–2050 and 3.6–5 °C for 2071–2100, and a decrease in rainfall by the end of the century. As a result, several direct and indirect impacts are expected in various sectors. Water resources will be increasingly stressed so that the already existing risk of desertification will increase.

© The Author(s) 2016
T. Zachariadis, *Climate Change in Cyprus*,
SpringerBriefs in Environmental Science, DOI 10.1007/978-3-319-29688-3_5

The biodiversity of the island may suffer from serious extinction of species and further invasion of alien species. In forests, the impacts may involve frequent forest mortality events and increased forest fire risk. In agriculture, a substantial portion of the island's crop and livestock production will be endangered. In public health, there will be a higher risk of emergence of specific diseases. Coastlines along the island are expected to experience serious degradation and sea water intrusion due to rising sea level. In the energy sector, additional power generation capacity will be needed in order to fulfil rising needs for space cooling during hotter summers and sea water desalination. The tourist sector may experience a significant loss of summer tourist arrivals due to increasingly inconvenient weather conditions during the hottest months of the year. There may also be impacts on the society in general: health deterioration will increase healthcare expenditures, increase insurance rates and affect labour productivity; adverse effects in various economic sectors may also lead to job losses.

The extent to which such effects will affect the society and economy of Cyprus depends crucially on policies to be implemented in the frame of a national adaptation strategy. As expressed clearly in the latest report of the Intergovernmental Panel on Climate Change and was reiterated at the agreement that came out from the global climate conference in Paris in December 2015,[1] mitigation and adaptation policies should be closely linked and can offer synergies for coping with climate change. For this purpose, proactive actions need to be taken by both the public and the private sectors. Public authorities need to set clear priorities and implement well-designed policies in line with international best practices and the recommendations outlined in this book. Most importantly, adequate monitoring mechanisms should be set up in order to provide much needed data which can send early warnings to policy makers and the public and can help avoid large natural and economic damages at a later stage. Enabling private adaptation investments and properly pricing the use of natural resources in the frame of a well-designed environmental fiscal reform are key priorities for investing in a climate resilient economy.

---

[1]See the full text of the Paris agreement at http://unfccc.int/resource/docs/2015/cop21/eng/109r01.pdf.

Appendix

List of potential climate change adaptation measures for Cyprus

Sector	Impacts (observed and expected)	Adaptation measures
Water resources	➤ Water scarcity	✓ Promote research, development and innovation
	➤ Declining quality of surface and groundwater	✓ Improved water management through proper implementation of the Water Framework Directive
	➤ Decline of aquifers	✓ Implement the adopted national strategy for sustainable development
	➤ Coastal aquifer salination	✓ Implement the management plans of the NATURA 2000 sites
	➤ Desertification	✓ Implement the action plan to combat desertification
	➤ Reduced crop yields	✓ Ecosystem-based adaptation-building ecosystem resilience
	➤ Biodiversity loss and ecosystem damage	✓ Sustainable use of ecosystem services and natural resources
		✓ Protection of the natural ecosystems especially protected areas and wetlands
		✓ Policies and measures to boost ecosystem water storage capacity
		✓ Maintain and restore wetlands and river beds as natural defence against floods
		✓ Waste management to avoid pollution and surface and groundwater deterioration
		✓ Raise public awareness
		✓ Maintenance and repair of the water distribution systems and related infrastructure (adoption of technologies for leakage detection and control)
		✓ Appropriate water pricing to reflect scarcity and to encourage efficient water use and conservation

© The Author(s) 2016
T. Zachariadis, *Climate Change in Cyprus*,
SpringerBriefs in Environmental Science, DOI 10.1007/978-3-319-29688-3

Sector	Impacts (observed and expected)	Adaptation measures
		✓ Sea water desalination as an ultimate solution and preferably by renewable energy
		✓ Control over the amount of the desalinated water produced in order to avoid disposal of costly desalinated water. Conventional desalination is energy intensive and is considered as maladaptation
		✓ Installation of water saving equipment
		✓ Adoption of efficient water consumption standards
		✓ Application of grey water treatment and recycling for indoor use and for irrigation
		✓ Mandatory grey water recycling for new houses/buildings
		✓ Collection and use of rainwater
		✓ Treated effluent reuse
		✓ Enhance water use efficiency in agriculture, households and buildings
		✓ Improvement of irrigation management (i.e. water supply, irrigation methods/systems and cultivation of local drought-resistant crops)
		✓ Prevent irrigation with brackish water
		✓ Promote soil management and good condition (avoid erosion, salination and uncontrolled use of fertilisers and pesticides) to maintain and enhance infiltration
		✓ Sound use of fertilisers and pesticides to protect surface and groundwater quality. Surface and groundwater quality monitoring
		✓ Monitoring and control of groundwater quantity. Sound use of groundwater—strict limitations where needed
		✓ Coastal aquifer replenishment
Ecosystems and biodiversity	➤ Changes in species phenology (timing of seasonal biological phenomena)	✓ Promote research on biodiversity and ecosystems, monitoring of biotic and abiotic parameters
	➤ Changes in species distribution	✓ Implement the adopted national strategy for sustainable development
	➤ Species extinction	✓ Implement NATURA 2000 management plans, considering also climate change implications
	➤ Reduced habitat availability	✓ Maintain or strengthen ecological coherence, primarily through providing for connectivity. Establishment of ecological networks (protected sites and corridors)
	➤ Endemic species are particularly vulnerable	✓ Prepare and implement a strategic plan on biodiversity
	➤ Invasive species	✓ Implement the adopted action plan to combat desertification

Sector	Impacts (observed and expected)	Adaptation measures
	➢ Algae blooms	✓ Implement-related provisions, included in the Rural Development Programme 2007–2013
		✓ Incorporate in other policies and plans [local plans, environmental impact assessment (EIA), strategic environmental assessment (SEA)] the priority of biodiversity and ecosystem protection in relation to climate change
		✓ Ecosystem-based adaptation-building ecosystem resilience
		✓ Horizontal integration of ecosystem-based adaptation to other policies and plans
		✓ Sustainable development policies land use
		✓ Enhance the preparation of the Island Plan
		✓ Sustainable use of ecosystem services and natural resources, particularly in areas of importance to biodiversity conservation
		✓ Full species inventory—populations, distribution, dispersal, genetics and determination of threatened species, monitoring
		✓ Special attention to protection of priority and threatened species and their habitats
		✓ Enhance/strengthen the seed bank and ex situ conservation
		✓ Monitoring especially the bioindicators and population- and species-based indicators
		✓ Highly sensitive species should be monitored as indicators of climate change, i.e. amphibians and reptiles
		✓ Installation of wildlife water supply system (also useful for forest fire protection)
		✓ Control overgrazing
		✓ Inventory of the alien species (plants/animals, terrestrial/marine) and their distribution
		✓ Avoid the planting of alien species and releasing of alien animal species
		✓ Avoid overfishing and any destructive fishing practices
		✓ Protection of coastal and marine ecosystems from invasive species (prevention–detection–control)
		✓ Restoration of damaged ecosystems (i.e. artificial dispersal of seeds, restore water bodies/flows, soil quality, remove alien species, etc.)

Sector	Impacts (observed and expected)	Adaptation measures
		✓ Assessment of the impacts of pollination disruptions on plant reproduction, protection of pollinators
		✓ Reduce pressure from agriculture. Reduce the impacts of fertilisers and pesticides/herbicides. Controlled use. Promote organic farming
		✓ Conservation of ecosystem functions
		✓ Waste management to avoid pollution, ecosystem degradation and surface and groundwater deterioration
		✓ Elimination of illegal dumping sites
		✓ Control over mining and quarrying activities
Forests	➤ Increased forest fire risk	✓ Research, data collection and monitoring of biotic and abiotic parameters
	➤ Increased frequency of forest mortality events-species necrosis	✓ Implement the national strategy for sustainable development
	➤ Species extinction	✓ Implement NATURA 2000 management plans, considering also climate change implications
	➤ Changes in distribution, species composition	✓ Maintaining or strengthening ecological coherence, primarily through providing for connectivity. Establishment of ecological networks (protected sites and corridors)
	➤ Changes in species phenology	✓ Implement the adopted action plan to combat desertification
	➤ Increased pest populations	✓ Implement/strengthen the provisions of the short-term action plan on combating the effects of drought on state forests (2009–2010)
	➤ Reduction in forest productivity—decline of wood biomass	✓ Implement-related provisions, included in the Rural Development Programme 2007–2013
	➤ Surface run-off and soil erosion	✓ Enhance the preparation of the Island Plan
	➤ Redistribution of species range due to contract and loss of habitats	✓ Ecosystem-based adaptation-building ecosystem resilience
		✓ Sustainable use of ecosystem services and natural resources
		✓ Raise public awareness (prevention, warning, volunteers)
		✓ Waste management to avoid pollution and surface and groundwater deterioration
		✓ Elimination of illegal dumping sites
		✓ Control over mining and quarrying activities
		✓ Protection of forest biodiversity (plants/animals)

Sector	Impacts (observed and expected)	Adaptation measures
		✓ Full species inventory—populations, distribution, dispersal, genetics and determination of threatened species, monitoring
		✓ Special attention to protection of priority and threatened species and their habitats
		✓ Enhance/strengthen the seed bank and ex situ conservation
		✓ Monitoring especially the bioindicators and population- and species-based indicators
		✓ Reforestation, afforestation
		✓ Identification and promotion of microclimatic benefits and environmental services of trees and forests
		✓ Infrastructure to improve forest resilience to fires
		✓ Promotion of sustainable forest management in order to reduce the fuel potential for forest fires (forest thinning)
		✓ Classification of forests according to the risk of fire, designation of high-risk areas
		✓ Early warning systems
		✓ Sustainable wood management
		✓ Irrigation systems/water supply systems for firefighting
		✓ Installation of watering devices for wildlife
		✓ Pest control
Agriculture and Livestock	➤ Acceleration in the loss of genetic and cultural diversity	✓ Implement CAP provisions related to climate change and desertification, suitable for Cyprus particularities
	➤ Water shortage, either for rain-fed and for irrigated crops	✓ Implement the adopted national strategy for sustainable development
	➤ Declined water quality	✓ Implement NATURA 2000 site management plans
	➤ Soil erosion	✓ Implement the adopted action plan to combat desertification
	➤ Reduced crop yields	✓ Implement the related provisions, included in the Rural Development Programme 2007–2013
	➤ Crop quality decrease	✓ Enhance the preparation of the Island Plan
	➤ Changes in the phenology of agricultural crops and in the growing season	✓ Ecosystem-based adaptation-building ecosystem resilience
	➤ Conflict over land use and other resources. Increasing competition for water between sectors and uses	✓ Sustainable use of ecosystem services and natural resources

Sector	Impacts (observed and expected)	Adaptation measures
	➤ Increased weed and pest challenges	✓ Sustainable use of water resources
	➤ Pest outbreaks, emergence of new pests and pathogens and increase in the frequency of diseases	✓ Develop infrastructure to harvest and store rainwater (especially for livestock and greenhouses)
	➤ Increased spread of existing vector-borne diseases and macroparasites, emergence and circulation of new diseases	✓ Improvement of irrigation management (e.g. through remote sensing)
	➤ Changes in the primary productivity of crops, forage, grassland, etc	✓ Use of treated effluents with strict monitoring to prevent build-up of emerging contaminants in soils
	➤ Heat stress suffered by live-stock animals will reduce the rate of feed intake resulting in poor growth	✓ Prevent overexploitation of groundwater
	Shortage of feed	✓ Implement the Code of Good Agricultural Practice, enhance organic and traditional agriculture
	➤ Declining of pollinating insect species	✓ Promote the sound use of fertilisers and organic manure (soil fertility management). Control/reduce the use of chemical fertilisers (especially in areas polluted with nitrates)
	➤ Increased concentrations of tropospheric ozone lead to decreases in plant biomass and yields	✓ Maintain and enhance soil fertility and residual soil moisture
		✓ Terracing, ridging, reduced tillage, deep ploughing etc., to retain soil moisture and organic matter and increase infiltration
		✓ Prevent soil salination
		✓ Prevent soil erosion and degradation (no-till agriculture, avoid the use of chemical herbicides, tree planting)
		✓ Promote agroforestry, examine silvo-pasto-ral systems
		✓ Control overgrazing
		✓ Reduce heat stress to livestock and improve crop yields by providing shade by planting trees. Use local species to enhance wildlife habitats (also positive effect on soil erosion prevention)
		✓ Enforce strict pesticide regulations and promote sound pesticide use

Sector	Impacts (observed and expected)	Adaptation measures
		✓ Changes in farming practices to avoid intensive chemical agriculture; crop rotation and fallow
		✓ Promote use of indigenous and locally adapted plants and animals as well as the selection and multiplication of crop varieties and autochthonous races adapted or resistant to adverse conditions:
		✓ Diversify livestock production. Identify and strengthen local breeds adapted to local climatic stress and feed
		✓ Avoid GMOs and monocultures, maintaining genetic diversity of crops/diversification of crops
		✓ The seed bank to cover all local varieties, ex situ conservation and special attention to in situ conservation/reintroduction of the local varieties
		✓ Select local heat- and drought-resistant crops
		✓ Promote local varieties seed exchange amongst farmers
		✓ Discourage the cultivation of high water-consuming crops or apply rotation with low water-consuming crops
		✓ Adjustment of sowing/planting season
		✓ Preserve/enhance family farming
		✓ Preserve and enhance diverse agricultural landscapes
		✓ Prevent rural depopulation
		✓ Conserve energy, use of renewables (on farm use of biomass where applicable and environmentally sound)
		✓ Support sound storage and processing of animal waste (convert to biogas)
		✓ Air pollution prevention (e.g. tropospheric ozone)
		✓ Protection of pollinators
		✓ Research and prevention of pest and disease outbreaks
		✓ Adjust animal husbandry methods
		✓ Introduce livestock management systems
		✓ Make use of historical data and local experience in coping with climate variability (certain livestock systems could more successfully adapt to climate change)

Sector	Impacts (observed and expected)	Adaptation measures
Human health	➤ Mortality due to higher temperatures	✓ Promote research, development and innovation
	➤ Mortality related to heat-waves and summer peaks	✓ Implement the adopted national strategy for sustainable development
	➤ Increased humidity–related mortality and morbidity	✓ Develop guidelines and proper training for medical doctors (private and public sector)
	➤ Declined air quality in urban areas—increased morbidity and mortality	✓ Implement an early warning system
	➤ Emergence of vector-borne diseases	✓ Data collection, inventory on vector, water- and food-borne diseases
	➤ Water- and food-borne disease outbreaks	✓ Monitoring and disease control
	➤ Enhanced allergies' risk	✓ Research in disease control and prevention
		✓ Establishment of general health scheme and horizontal integration of the climate change adaptation priority in all sectors
		✓ Improve health infrastructure (hospitals, laboratories, etc.)
		✓ Health- and social care systems must develop contingency plans to cope with increasing numbers of patients
		✓ An emergency plan should be prepared to specify the responsibilities of various health and social service bodies
		✓ Enhanced implementation of measures for air quality improvement in urban areas
		✓ Creation and protection of urban parks to reduce the urban heat island phenomenon and improve air quality
		✓ Apply biophilic and bioclimatic architecture for environmentally friendly, energy-efficient buildings by effectively managing natural resources, instead of using energy-intensive air conditioning which is a maladaptation
		✓ Monitor drinking water quality
		✓ Apply strict controls/health inspection in food industry and food service industry
		✓ Use international standards like ISO and HACCP
		✓ Raise public awareness
Coastal zones	➤ Inundation, flood and storm damage	✓ Promote research, development and innovation
	➤ Larger extreme waves and storm surges	✓ Research on sea-level rise. Increase monitoring sites and apply model simulations
	➤ Erosion–coast retreat	✓ Implement the adopted national strategy for sustainable development

Sector	Impacts (observed and expected)	Adaptation measures
	➢ Sea water intrusion and aquifer salination	✓ Implement NATURA 2000 management plans, considering also climate change implications
	➢ Coastal ecosystem degradation/loss	✓ Ecosystem-based adaptation-building ecosystem resilience
	➢ Wetlands dry out, or flooded because of sea-level rise	✓ Sustainable use of ecosystem services and natural resources
	➢ Infrastructure loss	✓ Integration of climate change implications into land use planning/local plans, EIA and SEA
	➢ Economic losses from erosion and coastal flooding	✓ Integration of adaptation in the implementation of Marine Strategy Framework Directive and in the Common Fisheries Policy
	➢ Damages to water resources	✓ Enhance the preparation of the Island Plan
	➢ People relocation	✓ Prepare an inventory of coastal areas already suffering from erosion as well as those vulnerable to erosion and evaluate the measures already taken
		✓ Examine the possibility of other measures to combat coastal erosion, since breakwaters consider maladaptation due to ecosystem/environmental damage they cause primarily in the quarrying areas
		✓ Protection of wetlands and sand dunes as a measure to combat erosion
		✓ Existing and new coastal developments/infrastructure should include adaptation measures to account for sea-level rise
		✓ Replenishment of coastal aquifers
		✓ Relocation of infrastructure and houses inland to allow coastal ecosystems to recover
Energy supply and demand	➢ Increased energy demand for cooling—higher energy expenditures for households and firms	✓ Promote research, development and innovation
	➢ Increased national electricity consumption due to sea water desalination	✓ Implement the adopted national strategy for sustainable development
	➢ Increased risk of electricity shortages in summer time	✓ Adaptation measures should not jeopardise the mitigation targets to reduce GHG, and on the other hand, the implementation of the EU climate and energy package should avoid maladaptation
	➢ Higher risk of power outages owing to insulator cleaning needs due to higher temperature and humidity	✓ Accelerated adoption of 'near-zero' energy new buildings, in conjunction with biophilic and bioclimatic architecture

Sector	Impacts (observed and expected)	Adaptation measures
	➤ More costly electricity production due to insufficient cooling of the generating units because of warmer sea water used as a cooling agent	✓ Investments in improving the energy efficiency of existing buildings
	➤ Additional/reserve capacity may need to be installed adding extra cost to energy production	✓ Adjustment of long-term electricity generation plans in order to account for additional capacity needed. Priority to be given to renewable electricity generation
	➤ The efficiency of photovoltaic plants could be slightly reduced by higher temperatures, especially during heat waves	✓ Implementation of proper carbon pricing of all energy forms in order to encourage energy conservation
	➤ Extreme weather events could cause power outages	✓ Proper maintenance of electricity transmission lines
		✓ Decide on scientific grounds the proper energy mix for Cyprus to avoid maladaptation
		✓ Raise awareness for energy saving (i.e. energy-saving appliances, controlled use of air conditioning)
		✓ Greening of towns to avoid/reduce urban heat island phenomenon, aiming to decrease energy consumption for cooling
Tourism	➤ Shift of tourist destinations further north and up mountains	✓ Promote research, development and innovation
	➤ Gradual decrease in summer tourism	✓ Implement the adopted national strategy for sustainable development
	➤ Indirect impacts due to:	✓ Implement NATURA 2000 management plans, considering also climate change implications
	➤ Coastal flooding and erosion that will cause damages to infrastructure	✓ Implement the adopted action plan to combat desertification
	➤ Biodiversity loss and ecosystem deterioration	✓ Ecosystem-based adaptation-building ecosystem resilience
	➤ Emergence of water-, food- and vector-borne diseases	✓ Sustainable use of ecosystem services and natural resources
	➤ Changes in air quality	✓ Enhance the preparation of the Island Plan
	➤ Increased fire risk	✓ Integration of climate change implications into land use planning/local plans, EIA and SEA
	➤ Extreme weather events	✓ Protect the natural environment
	➤ Declining water quality and coastal erosion reduce the attractiveness of tourist sites	✓ Re-examine past decisions to permit the construction of 14 golf courses

Sector	Impacts (observed and expected)	Adaptation measures
	➤ Increased heat stress, more variable weather and precipitation	✓ Raise public awareness, education and training
		✓ Develop guidelines for sustainable tourism and adaptation to climate change
		✓ The tourism industry should rapidly respond to the expected decrease in summer tourism and to the shift in tourist distribution to spring and autumn
		✓ Aim at a more uniform distribution of tourist arrivals over months/seasons
		✓ The tourism industry should take action to combat the emerged competitiveness of other destinations in Europe, which will be favoured by climate change
		✓ Investments in infrastructure/technologies to upgrade facilities to face increased temperature and water shortage
		✓ Measures to counteract possible weather extremes and flooding
		✓ Shift from mass/coastal tourism to special interest tourism (i.e. ecotourism, agrotourism)
		✓ Investments to reduce climate change/ carbon footprint. Gain competitive advantage from such actions
		✓ Water conservation, efficient use and reuse. Manage/reduce tourism's water footprint
Social impacts	➤ Reduced security due to weather extremes and fires	✓ Implement the adopted national strategy for sustainable development
	➤ Increased healthcare expenditures	✓ Society must get prepared to pay for climate change adaptation measures
	➤ Changes in mortality and morbidity, impacts on labour productivity	✓ Increase education and training. Raise public awareness and mobilise society to change consumption patterns and avoid overconsumption of resources
	➤ Changes in economic sectors (agriculture, fisheries, tourism)—job losses	✓ Allocate funds to enable low-income households to afford capital expenditures for adaptation/mitigation measures, in order to avoid 'energy and climate poverty'
	➤ Reduced recreational services demand due to changes in tourism distribution	✓ Social security services need to prepare a relevant strategy
	➤ Water shortage—declined standard of living because of water supply limitations	✓ Be prepared for a possibly increased numbers of environmental refugees

Sector	Impacts (observed and expected)	Adaptation measures
	➤ Loss of profit due to ecosystem damage	✓ Cypriot insurance should consider the development of new insurance products to cover climate change impact risks
	➤ Decline in crop yields and livestock production. Reduced GDP from agriculture	
	➤ Capital losses due to infrastructure damages	
	➤ Coastal land loss due to inundation and erosion	
	➤ Capital losses due to infrastructure and people relocation	
	➤ Increased insurance rates	
	➤ Migration and civil unrest	
	➤ Receiving and care of environmental refugees	

Printed in the United States
By Bookmasters